KU-682-925

PRAISE FOR *DRIVE AND CURIOSITY*

"Istvan Hargittai shares with us his fascinating insights into the passion and drive that lie behind some of the most important scientific discoveries of the last century. This is an invaluable book for those who want to go beyond the usual textbook descriptions of the scientific endeavor."

—Leon N Cooper, Nobel laureate,
Brown University

"Hargittai has succeeded in summarizing in a very skillful manner the lives of many of the most important scientists of the last half of the twentieth century. His seemingly casual style is adept in portraying the personalities and motives of these scientists very vividly."

—Sidney Altman, Nobel laureate,
Yale University

"Yet another masterpiece by Istvan Hargittai, this time highlighting the multifarious patterns of human endeavor to be found behind some of the most fascinating and astounding chapters in the history of science."

—Arvid Carlsson, Nobel laureate,
University of Gothenburg, Sweden

"This book presents insights into the intricacies of scientific research and discoveries and should find wide interest outside as well as inside the scientific community. The stories can in particular motivate young people to become scientists. This stimulating book will greatly help the reader to understand and to appreciate scientific advances as well as the impact of their practical applications on our living conditions."

—Werner Arber, Nobel laureate,
University of Basel, Switzerland

"The eternal quest for creativity assumes many paths. Hargittai's book has a collection of unusual paths that led to successful solutions for existing problems and also shed light upon unexpected new knowledge that allows us to understand our universe better. The book provides approaches that have changed 'fuzzy logic' to concrete physical or mathematical relationships."

—Isabella Karle, formerly of the
Naval Research Laboratory;
recipient of the National Medal of Science

"Science, discovery, and scientists: the right person, at the right time, in the right place. This book provides a thoughtful and vivid analysis of how scientific progress happens through some selected case studies. It should be of value to the practitioners but even much more so to the general educated public who wish to get insight into the various processes underlying the quest for knowledge. Prometheus gave us the light and we cannot give it back. Science will shape our destiny!"

—Jean-Marie Lehn, Nobel laureate, France

"Hargittai's book offers potential explanations of the origin of the most laudable of all human endeavors—the pursuit of high-quality science. This is done by describing the paradigmatic achievements of fifteen selected scientists, most of whom have been interviewed personally by the author. The stories are exciting to read and they illustrate that a wide diversity of personal traits may further the making of major discoveries. There is no predictable single road to success."

—Erling Norrby, former dean of the
medical faculty of the Karolinska Institute,
member of the Nobel Committee for
Physiology or Medicine, permanent secretary
of the Royal Swedish Academy of Sciences,
and member of the board of the
Nobel Foundation, Stockholm

"This new book by István Hargittai addresses a theme of great interest. Not only parents, students, and young scientists are keen to know the recipe for a successful career in science, but also mature scientists themselves and science historians. . . . The book answers this question in an indirect way, in a series of lessons, consisting of fifteen highly interesting stories of major scientific discoveries. Each illustrates a special trait shown, in addition to drive and curiosity, by the scientists making the discoveries. In this way Professor Hargittai manages to touch on such useful traits as bravery, dedication, stamina, toughness, aggressiveness, perseverance, serendipity, risk-taking, imagination, and many more. Among the twenty scientists who are the main players in these stories, more than half have so far received Nobel Prizes. A few have passed away without winning, but one or two may still be given this very particular distinction. If so, Professor Hargittai can really be said to have succeeded in his choice of the corresponding lessons!"

—Anders Bárány, former professor of physics,
scientific secretary to the Nobel Committee
for Physics, and deputy director
of the Nobel Museum, Stockholm

"Drive and curiosity—the power behind the relentless human will to build, explore, invent, understand the earth and the universe, unveil the secrets of nature, and carve new ways to make life better and nicer—are hard to define but easy to identify once seen. Along with a few other ill-defined, vague, and volatile attributes, such as experience, serendipity, timing, and mentorship, and, yes, somewhere among them also wisdom and intelligence, they constitute the infinitely thin trunk that carries on it the entire tree of human achievements along history. Science, a young discipline, has joined this legendary list of historical achievements only recently. Istvan Hargittai has collected for us some of the most important scientific discoveries of the twentieth century—from the secret of creation through that of genetic inheritance, and from understanding the materials that surround us through our ability to look into ourselves in a magic, noninvasive manner. All these majestic discoveries were made by brilliant people who did not have any detailed plan before them, paradigms to follow, models to imitate, or recipes to copy. Furthermore, at the time they made their discoveries they did not even have the ability to understand the full, or even the partial scope of their significance. These discoveries all stemmed from this mystique stew of vague human characteristics—drive and curiosity first and foremost.

—Aaron Ciechanover, Nobel laureate,
TECHNION Medical School, Haifa, Israel

DRIVE
AND CURIOSITY

ALSO BY THE AUTHOR

Judging Edward Teller: A Closer Look at One of the Most Influential Scientists of the Twentieth Century (Prometheus Books, 2010).

With Magdolna Hargittai, *Symmetry through the Eyes of a Chemist*, 3rd ed. (Springer, 2009; 2010).

With Magdolna Hargittai, *Visual Symmetry* (World Scientific, 2009).

The DNA Doctor: Candid Conversations with James D. Watson (World Scientific, 2007).

The Martians of Science: Five Physicists Who Changed the Twentieth Century (Oxford University Press, 2006; 2008).

Our Lives: Encounters of a Scientist (Akadémiai Kiadó, 2004).

The Road to Stockholm: Nobel Prizes, Science, and Scientists (Oxford University Press, 2002; 2003).

With Magdolna Hargittai and Balazs Hargittai, *Candid Science I–VI: Conversations with Famous Scientists* (Imperial College Press, 2000–2006).

With Magdolna Hargittai, *In Our Own Image: Personal Symmetry in Discovery* (Plenum/Kluwer, 2000).

With Magdolna Hargittai, *Symmetry: A Unifying Concept* (Shelter Publications, 1994).

With R. J. Gillespie, *The VSEPR Model of Molecular Geometry* (Allyn & Bacon, 1991; Dover Publications, 2012).

ISTVAN HARGITTAI

Foreword by Carl Djerassi

DRIVE AND CURIOSITY

What Fuels the Passion for Science

CHARTERHOUSE LIBRARY

WITHDRAWN

Prometheus Books

59 John Glenn Drive
Amherst, New York 14228–2119

Published 2011 by Prometheus Books

Drive and Curiosity: What Fuels the Passion for Science. Copyright © 2011 by Istvan Hargittai. All rights reserved. No part of this publication may be reproduced, stored in a retrieval system, or transmitted in any form or by any means, digital, electronic, mechanical, photocopying, recording, or otherwise, or conveyed via the Internet or a website without prior written permission of the publisher, except in the case of brief quotations embodied in critical articles and reviews.

Cover image © Media Bakery, Inc.
Jacket design by Liz Scinta

Inquiries should be addressed to

Prometheus Books
59 John Glenn Drive
Amherst, New York 14228–2119
VOICE: 716–691–0133
FAX: 716–691–0137
WWW.PROMETHEUSBOOKS.COM

15 14 13 12 11   5 4 3 2 1

Library of Congress Cataloging-in-Publication Data

Hargittai, Istvan.
    Drive and curiosity : what fuels the passion for science / by Istvan Hargittai.
        p. cm.
    Includes bibliographical references and index.
    ISBN 978–1–61614–468–5 (cloth : alk. paper)
    ISBN 978–1–61614–469–2 (ebook)
    1. Scientists—Intellectual life. 2. Discoveries in science. I. Title.

Q147.H37 2011
509.2'2—dc23

                                                              2011023753

Printed in the United States of America on acid-free paper

For
Magdi
Eszter
Balazs and Michele
Matthew and Stephanie

"Lonely Discoverer." Drawn by and courtesy of professor of mathematics Anatoly T. Fomenko, Moscow State University. Used with permission.

CONTENTS

FOREWORD

Carl Djerassi

*D*rive and Curiosity is an unusual book by an unusual author. Istvan Hargittai is a working scientist who is also a prolific author of a special genre of books, which—directly or indirectly—describe the nature of *Homo scientificus*. Most of his conclusions are based on direct interviews conducted in a nonintrusive yet searching manner that instills confidence and frequently leads to disclosures on the part of the interviewee that one would not have expected. I speak from personal knowledge, having gone years ago through such a Hargittai interrogation, which prompted me to search in some of his other books for further examples among famous or notorious scientists of confessions, mea culpas, instances of braggadocio, and on occasion even aspects of automythology. The latter is not surprising, since so many autobiographical musings are by definition tainted by automythology as they pass through a person's psychic filter.

In this latest book, Hargittai has collected over a dozen interviews (amplified by literature research) with actual or quasi- (i.e., nonanointed) Nobel laureates, which he has woven into an intriguing account of how great scientific discoveries are made. As the title implies, he assigns *drive* and *curiosity* the biggest roles, to which I would have added *serendipity*—given that in several of his examples sheer good luck was the crucial component. Among sci-

9

entists, *curiosity* is a given and *serendipity* an unsolicited gift. But what does *drive* mean? It is here that the reader will find intriguing examples, ranging from strict lifelong workaholic discipline to working modes that border on dillydallying. How could *drive* cover such a broad range? Quite simply because in every case *ambition* is the key component—the motivation of so many scientists, but at times also their poison.

Hargittai's book has two remarkable features. In fifteen chapters, he manages to cover an extraordinary range of scientific fields—all of them presented in a style that is attractive to readers ranging from sophisticated scientists to laypersons. In the process, he has created a book that may well influence young women and men to select some scientific discipline for their ultimate profession. What makes me say that?

In the author's preface, Hargittai states: "From my numerous inquiries, I learned that an inspiring book has turned more children to science than any other single ingredient such as a teacher or a member or friend of the family." I share that opinion and still recall asking that question of the Nobel laureate Joshua Lederberg and hearing him cite Paul de Kruif's *Microbe Hunters* as such a book—a choice that many scientists of my generation (including myself) would have offered, although these days it would seem outmodedly romantic. Apparently, it was also Gertrude Elion's favorite. I would not be surprised if Hargittai's *Drive and Curiosity* will join this list of seminal books.

Hargittai avoids hagiography when describing the work and personalities of the two women and seventeen men that he selected as examples for his generalizations of what made them tick as scientists. With one exception—Kary Mullis—I think that all others illustrate in a positive sense the range of behavioral aspects of *Homo scientificus* from which prospective scientists can learn something constructive. In my opinion, Kary Mullis illustrates none of them except for the operation of serendipity and excessive automythology, coupled with a startling unwillingness to acknowledge the crucial role of colleagues. That the polymerase chain reaction (PCR) is spectacularly important and clearly deserving of a Nobel Prize is unquestionable; I myself nominated him on two occasions for the Nobel Prize, though both times with two collaborators who converted the idea of PCR into practical reality. But much of what is recorded in chapter 11 about his discovery and his behavior is what I'd consider self-sanitized semifiction. Hargittai's description "he appears gentle and diffident, with a sense of self-deprecating humor" may well be true now, but it certainly was not a quarter of a century ago when the discovery was made. The statement "the Nobel Prize paved his way to publish his first and so far only book" is also true, but what about that book, titled *Dancing Naked in the Mind Field*? It is essentially a potpourri of autohagiographic ramblings about abductions by outer space aliens, astral travels with a lover, Mullis's belief in the innocence of O. J. Simpson, his conviction of a lack of relation between HIV and AIDS, and his judgment that the impending destruction of the ozone layer is a plot by DuPont rather than the outcome of research by two Nobelists actually featured in chapter 10 of Hargittai's book. I can think of no worse example as a model for a new generation of scientists who ought to be inspired by drive and curiosity.

But the other fourteen chapters more than make up for this controversial pick. Watson, Pauling, Sanger, and probably also Teller are obvious and yet diverse choices that illustrate different aspects of *drive*. Neil Bartlett's is a particularly felicitous selection in demonstrating that not every "paradigmatic" discovery ends up with a Nobel Prize. (I liked that Hargittai remembered Primo Levi's understandable error in the first chapter of his *Periodic Table* of assuming that Bartlett had won a Nobel Prize for his discovery that noble gases could also be made reactive.) The inclusion of the two women Nobelists, Gertrude Elion and Rosalyn Yalow, is especially appropriate since it illustrates the extraordinary hurdles that women faced in science only a few decades ago. Yet instead of glamorizing them solely as heroines, Hargittai shows their strengths, sacrifices, and monumental drive—and in Yalow's case also blemishes.

The inclusion of Rowland and Molina is instructive in many regards, but especially in terms of demonstrating how *persistence* —another component of *drive*—can ultimately lead to globally implemented policy changes. Although not explicitly stated, their work on atmospheric ozone depletion is probably one of the best counterarguments to today's chemophobia: by definition, chemicals are the cause of various environmental and toxicological problems, but chemistry is also the answer for combating most of them.

Two Hungarian representatives, Árpád Furka and Leo Szilard, merit special attention. To me, Furka came as a total surprise since I had never heard of him, always having associated his crucial methodological advance first with Bruce Merrifield and subsequently with Mario Geysen. Chapter 6, focusing on Furka, convincingly demonstrates that even scientists suffering under the delusion that they are well-informed on combinatorial chemistry can learn something new and important in this book.

But dedicating an entire chapter to Leo Szilard seemed to me truly inspirational, since it also covers an aspect of *Homo scientificus* that is usually ignored; namely social and even political engagement. Although I have never met Szilard personally, he was always a hero in my mind. Intellectually, he was truly astounding, and Hargittai explains succinctly his enormous scientific contributions in physics and to a lesser extent in biophysics. Yet he won few major scientific awards, because he was primarily a planter of intellectual ideas in many fields rather than a consummator in any one of them. But more than most twentieth-century scientists, he understood the global consequences of some of the ideas to which he contributed so deeply—first by pushing for the development of an atomic bomb, but then by dissuading people from using it. He was one of the first participants in the Pugwash movement as well as a founder of the Council for a Livable World, which primarily offers political support to senators who are prepared to concern themselves with arms control (with a focus on small states—the argument being that a senator in a small state, where political campaigns are less costly, has the same vote as one in a large state; after all, only in the Senate are Rhode Island and Texas the same size). I am mentioning these Szilardian activities because Szilard selected a unique method to present and propagate his ideas to a general public, namely in the guise of fiction. In my opinion, his short story *The Voice of the Dolphins* ought to be compulsory reading on litcrary, humorous, and most importantly political grounds. Briefly, his 1961 short story postulated that during the Cold War, the Americans and Russians decided to collaborate in a neutral site—specifically Vienna—on a project of nonmilitary significance. Founding the Vienna Biological Research Institute with a focus on studying the intelligence of dolphins, they developed AMRUSS—a combined contraceptive and foodstuff, which made the institute financially independent. I shall not monopolize this space so generously offered by Istvan Hargittai to give away the manner in which the institute spent its enormous financial resources in the political arena, other than to mention that it bore an eerie resemblance to actual events of that same period: the founding in 1972 in Laxenburg near Vienna of IIASA (International Institute for Applied Systems Analysis), which used the then-emerging discipline of systems analysis as a medium for peaceful engagement with the Soviet Union. I have often thought that subliminally, Szilard's *The Voice of the Dolphins and Other Stories* eventually influenced me to choose the genre of "science-in-fiction" to propagate some of my own ideas about the behavior and culture of scientists in a belletristic fashion.

The remaining Hargittai choices reach far and wide in terms of subject matter (MRI, X-ray crystallography, conducting polymers, and cosmology), but what I find most impressive is that they—as well as the rest of the book—illustrate the enormous range of work habits of twentieth-century scientists. His summary—*the best one can do is doing what one is best at doing*—also applies in spades to his contribution as a writer on the nature of *Homo scientificus*.

Carl Djerassi,
professor of chemistry emeritus, Stanford University;
recipient of the first Wolf Prize in Chemistry,
the National Medal of Science,
and the National Medal of Technology

PREFACE

by Harold Kroto

Fifteen fascinating scientific discoveries, all but one of which were made in the second half of the twentieth century, are examined in Istvan Hargittai's new book. We learn about the circumstances of the discoveries and also something about the individuals who made them. Although the way the universe works is fundamentally independent of the human spirit, the way that knowledge is uncovered and the way we describe it and use it is not. The mix of knowledge and personal information about individuals and their discoveries that Hargittai presents gives us a deeper understanding of science as a human endeavor and explains why the enthusiasm for uncovering new understanding continues unabated. The book benefits greatly from the fact that not only is the author a practicing scientist with great expertise in a broad portfolio of research areas; he has also been closely associated with all but two of the scientists to whom he introduces us. A further merit is the economical and engaging style of writing, which benefits greatly from his personal knowledge.

The book covers the daring decision of Watson and Crick to embark on a quest, which many judged premature or even impossible, that resulted, eventually, in the elucidation of the human genome. Watson's flamboyant style is contrasted with the natural modesty of Fred Sanger, whose breakthroughs ultimately made the elucidation of the human genome possible. In the story of the development of MRI, we see the struggle for the acquisition of knowledge of the scientists who were responsible for the tremendous benefits MRI now offers as a diagnostic tool. Hargittai introduces us to the additional hurdles that two remarkable women had to overcome on their way to developing new medical approaches. These are but a few of the examples covered in the book. Along the way he explores the individual variations in the complex mix of curiosity, stubbornness, perseverance, risk taking, and competitiveness, as well as the other human characteristics that underpin the road to discovery.

Istvan Hargittai has that most valuable quality, in the context of this treatise, a fascination with the personal details involved in important discoveries. His aim is to shed further light on the way scientists work as he uncovers details that are not contained in the original papers and also often not in the personal accounts of the scientists themselves. Because he writes from his own personal perspective and conflates his work with that of the primary authors, we are able to develop a greater overall understanding of the dynamics of the discovery process. This is of value as far as the history of science is concerned, and it is also particularly useful for students who can learn that there are as many ways to do science as there are scientists. In my experience, there is a wide spectrum of general approaches: at one end are the scientists who deliberately set out to solve highly challenging problems, and some are successful—though many, of course, are not; at the other end are scientists who go their own way, exploring problems they find personally curious, oblivious to whether or not the problem seems important to others, or whether there is or is not competition. This latter group, if they do make major breakthroughs, do so unexpectedly and serendipitously. Of course, this cohort includes many who do not make major breakthroughs, but it does tend to include many scientists who are relatively satisfied with their lives and at one with their achievements. The former cohort tends to contain a rather large number of rather disappointed individuals!

Scientists make up less than 0.2 percent of society, even in the developed world, and yet this 0.2 percent is basically responsible for the fundamental discoveries that underpin the technological world we now live in. It is vital that more people understand science not just as a cash cow only to be milked so industry can make money or even just for the social benefits that might possibly be predicted to accrue. It is vital that the other 99.8 percent of society that benefits from science understands how vital unpredictable and serendipitous discoveries have been and, most importantly, how crucial cultural aspects have been to the science underpinning technological paradigm shifts. The survival of the human race depends on this today more than ever before. Perhaps it is most important that politicians, who daily make decisions on technological issues, understand the most important aspects of the way science works. After all, it should be clear to anyone that though knowledge cannot guarantee good decision making, common sense suggests that wisdom is an unlikely consequence of ignorance.

If this text makes some headway in enhancing the appreciation of science as a cultural activity, it will have done much to improve matters. Hargittai has chosen his examples with care and has astutely covered a broad range of areas from biomedicine to cosmology, and the perpetrators are a richly diverse bunch of characters. In general the discoveries chosen have profoundly impacted our lives, and the discoverers are always uniquely interesting though perhaps not all are the best of role models. This book is a good read, and scientists and nonscientists will enjoy it and find in it much fascinating, unexpected, and thought-provoking information. Hargittai does indeed achieve his aim, which is to add greatly to our understanding of science as a uniquely human cultural activity.

Sir Harold "Harry" Kroto, Nobel laureate,
Francis Eppes Professor at Florida State University

INTRODUCTION

by Robert Curl

*D*rive and Curiosity shows great science done by individuals of completely different personalities and backgrounds. We learn from it that you can make a huge contribution to the world even if you spent your early career apparently adrift. We learn that you can be filled with passion for your research or apparently be always relaxed and calm and succeed. We learn that great advances in human knowledge often are the result of happy accidents but can also be the result of years of backbreaking toil. We see individuals who come close to the traditional image of the scientist as being shy, wishing to avoid the spotlight belonging to the group. And we find others who are more like rock stars, and still others involved intensely in communication, trying to shape government policy. We follow a scientist trying to make a great discovery where literally the fate of the world hangs upon his success . . . and succeeding.

But this book is not just about individuals; it describes how science really works as the vast social enterprise that it is. We see some individuals not getting the credit they deserve until well after their death. We see ethical guidelines crossed by truly great scientists . . . once just to create a joke. We see the dilemma of the young scientist who realizes the advantages of working with a person of great reputation but also realizes that if he happens to be part of a great discovery the community will credit the famous man. We begin to understand how this concern applies with redoubled force if the young scientist happens to be a woman. We become acquainted with women who overcame enormous hurdles to win through to great success and fame.

Drive and Curiosity shows the scientific community punishing individuals because of the notoriety resulting from exposure of private behavior. We see scientists being punished for discovering and reporting unpleasant facts that require a change in the world's behavior. The story of stratospheric ozone depletion is being replayed with a vengeance in the current struggle over global warning. With no easy fix such as that found for ozone depletion, we see scientists being vilified. It has proved much easier to attack the scientist than to attack the science.

Drive and Curiosity depicts scientists as real people working in the real world of science, not as unreal people in the fantasy world imagined by the entertainment industry. There inevitably is a lot of science in it. The reader may not be familiar with the science, but it is so clearly explained that the reader is likely to be encouraged to learn more about it. If not, it is a good read anyway. Enjoy!

Robert F. Curl, Nobel laureate,
K. S. Pitzer-Schlumberger
Professor of Natural Sciences Emeritus, Rice University

AUTHOR'S PREFACE

I have been involved with scientific research since my work for a master's degree during the 1964–1965 academic year, but my interest in how other scientists make discoveries developed only thirty years later. The question that intrigued me was about the origin of scientists' interest in going into research. From my numerous inquiries, I learned that an inspiring book has turned more children to science than any other single ingredient, such as a teacher or a member or friend of the family. My interest was also sparked by a book; it was about chemistry, and I received it when I was ten years old.[1] However, the time when one decides to become a researcher comes much later, usually during graduate studies. Working for a successful scientist is probably the best route to becoming one.

There is no exact definition of what a scientific discovery is, except that it must be something new. The saying is attributed to Enrico Fermi that if one makes an experiment and gets what was expected, it is a measurement. If one gets something different from what was expected, it is a discovery. My students usually like this definition, although their enthusiasm is dampened when they realize that the reason for getting something different from what is expected may also be some error in the course of the experiment. I like Albert Szent-Györgyi's definition, according to which one makes a discovery when one sees what everybody else does but thinks what nobody had thought before.[2]

It is seldom possible to follow the process of discovery in all its depth, because the journals in which discoveries are reported frown upon descriptions about details that are deemed nonessential and upon historical narratives. Besides, the true significance of discoveries seldom appears obvious in the initial publication. Subsequently, the discoverers may find it hard to recollect the events leading to the discovery. Different participants in the same discovery remember it differently even in the most benevolent cases. Furthermore, the discoverers may choose to gloss over blind alleys and other difficulties or may just simplify the story to make it more accessible to their audience.

Following the process of a discovery can be a source of excitement and intellectual pleasure. Discussing the path to scientific discoveries can bring together scientists and nonscientists and help bridge the gap between C. P. Snow's "Two Cultures." I have taught courses of science for nonscience majors and gave my first such course in 1986 at the University of Texas at Austin. Initially I was not sure whether this was an important course. Then someone told me that this might be the most important course I would ever give: introducing future lawyers, politicians, journalists, and economists to science might have a greater impact on society than teaching chemistry or physics to future scientists.

In 1999, I initiated a course at the Budapest University of Technology and Economics about great discoveries in the twentieth century. It was about the nature of discoveries in science and built on the hundreds of conversations I recorded with eminent scientists during the last decade of the twentieth century and the first years of the twenty-first century. Over two hundred of these interviews appeared in the six-volume book series *Candid Science*.[3]

In addition, I have given numerous talks about the nature of scientific discovery at various gatherings, including some at which the audience consisted of the most active researchers. An example was when Richard Lerner of the Scripps Research Institute, in La Jolla, California, asked me to give a talk at the meeting "Frontiers in Biomedical Research" in February 1998 in Indian Wells, California, and another was when Ingemar Ernberg of the Karolinska Institute, Stockholm, asked me to give a talk at the second annual retreat of the Karolinska cancer researchers in September 2003 on the island of Sandhamn in the Stockholm Archipelago. Other talks to scientists have been augmented with evening conversations in the company of friends and acquaintances at beautiful locations such as the French Riviera. On that particular occasion, a select group of people of the most diverse backgrounds came together and spontaneously talked at length about discoveries.

Both formal teaching and informal chatting encouraged me to write this book. It is often said that the general public has become alienated from science and views it with suspicion. This may well be the result of lack of knowledge, for which scientists may also be blamed just as anybody else: the general public by itself can do very little about becoming better versed in science, but scientists can do a lot more to promote and explain it. This book aims to decrease the gap between those who make scientific discoveries and those who would benefit from knowing about these discoveries.

It is my hope that nonscientists will find the book to be an accessible account of scientific discoveries and that scientists will read it to become informed about discoveries in fields that are not their own. Even scientists who are involved in a research field related to a particular chapter might find it of interest for facts and stories hardly known by others than the discoverers themselves. Much of the information and stories described here have come directly from the discoverers themselves; my role is that of an interpreter.

In most of the following fifteen chapters I describe discoveries whose main players I got to meet and to know to some extent. The only two whom I never met were Leo Szilard and George Gamow, two of the most colorful personalities of twentieth-century science; I feel the absence of this personal experience especially keenly. I have had several meetings during the past decade with James D. Watson (Jim) and his wife, Liz, including a stay at Cold Spring Harbor, New York, for three months, where my wife Magdi and I were their personal guests, and also in Budapest, Hungary; Cambridge, England; Woods Hole, Massachusetts; and New York City. We visited with Gertrude Elion at Research Triangle Park, North Carolina, and continued our relationship through correspondence.

In addition to visits to the homes and laboratories of Paul Lauterbur, Peter Mansfield, Alan MacDiarmid, Alan Heeger, Sherwood Rowland, and Kary and Nancy Mullis, we had additional opportunities to chat with them in Lindau, Germany, the site of gatherings of Nobel laureates. I have met with Frederick Sanger several times, both in Cambridge and at his home nearby. In addition to visiting Rosalyn Yalow, my acquaintance with Eugene Straus—her closest associate after Solomon Berson's death—augmented my personal experience with her. I have enjoyed a long friendship and many get-togethers with Árpád Furka and Danny Shechtman. In addition to my personal meetings with Neil Bartlett, our correspondence gave me a glimpse into his personality. Of the personal interactions, Edward Teller stands out not only because of our meeting and later correspondence but also because my recent book about him followed two years of studying his life and oeuvre. Each and every one of these scientists deserves my gratitude for selflessly sharing with me their thoughts and worries, experiences and opinions; moreover, my encounters with them served as an additional education for me. My own research has brought me to areas of physical chemistry over the years that overlapped to a small extent with various aspects of the works of Linus Pauling, Árpád Furka, Dan Shechtman, Alan MacDiarmid, and Neil Bartlett.

I would like to express my thanks to those colleagues and friends who read one or more chapters in draft at various stages of their readiness and gave me their comments and suggestions. They include Richard Ernst (chapter 3), Árpád Furka (chapter 6), Igor and Elfriede Gamow (chapter 15), Alexander Gann (chapter 1), Richard Garwin (chapter 14), Ronald J. Gillespie (chapter 12), Boris S. Gorobets (chapter 15), Sándor Görög (chapter 2), Richard Henderson (chapter 5), Zelek Herman (chapter 4), György Inzelt (chapter 9), Alexej Jerschow (chapter 3), William Lanouette (chapter 13), Peter Mansfield (chapter 3), Kary and Nancy Mullis (chapter 11), F. Sherwood Rowland (chapter 10), Nadrian C. Seeman (chapter 11), Marjorie Senechal (chapter 8), Dan Shechtman (chapter 8), Eugene Straus (chapters 2 and 7), Elizabeth Watson (chapter 1), John Waugh (chapter 3), Kurt Wüthrich (chapter 3), and Balazs Hargittai who read and commented on the entire manuscript. I found their criticism useful and their suggestions instructive, and I followed through on most of them. My appreciation to these reviewers does not preclude my responsibility for everything that is in this book.

I am grateful to the Budapest University of Technology and Economics and to the Hungarian Academy of Sciences for various opportunities that contributed to the creation of this book. This work is connected to the scientific program of the "Development of quality-oriented and harmonized R+D+I strategy and functional model at BME" project and was supported by the New Széchenyi Plan (Project ID: TÁMOP-4.2.1/B-09/1/KMR-2010-0002). I thank Ms. Judit Szűcs and Dr. Zoltán Varga for dedicated technical assistance.

I express my gratitude to Linda Greenspan Regan, executive editor at Prometheus Books, for her unfailing editorial guidance.

There is then one principal person behind all my efforts: my wife, Magdi—my friend and partner in all my adventures.

AUTHOR'S INTRODUCTION

I was driven.
  Peter Mansfield[1]

I am a scientist in large part because I was born highly curious.
  James D. Watson[2]

All scientists claim having curiosity; few admit to possessing drive. It is easier to define curiosity and harder to define drive. Yet there are often so many obstacles in a scientific career that, without being driven, it would be difficult to satisfy curiosity. Scientists seldom restrict their working hours from nine to five even under ideal conditions. Competitiveness in science is more widespread than many would believe and necessitates further efforts. Lately, scientific research has become an industry, but in our discussion we are mainly concerned with the kind of scientists who are dedicated to making discoveries, and not the growing masses of scientific workers.

Drive and curiosity do not always yield discovery, and when they do, the discovery is usually a minor one. We will be concerned here with milestone discoveries; in reality, however, the various characteristics and invested efforts are similar whether they result in a milestone discovery or in a minor find.

The discoverer often must endure loneliness, because a discoverer possesses something no one else does for a period, whether short or long. The discoverer is willing to stick his or her neck out of the crowd, because the discovery is not only always new; it is often even contrary to previous beliefs. There may be barriers to overcome, and such barriers are often erected by other researchers who might be more knowledgeable than the discoverer.

It often happens that the discoveries are made by scientists who are not among the greatest in their fields. Great scientists know more about their branch of science than most of their peers. This knowledge may also hinder their imagination. It is tempting to assume that something utterly new could not be right, and the great scientist knows every reason why it could not be right.

There is a difference between a genius and a great scientist. The genius recognizes connections between facts, phenomena, theories, and observations that had not been connected before in other people's minds. This manner of thinking might even be helped by some ignorance. Sometimes it is beneficial not to know about some limitations because such knowledge might curb one's imagination.

Among every one of the scientists in this volume, both drive and curiosity fuel their passion. And then there is an additional trait especially characteristic of the discoverer. This additional, more unique trait will be singled out for every individual discovery among the fifteen cases.

By learning about these discoverers and discoveries, we might come to a conclusion that would be applicable to more than just scientific research, a conclusion that might enrich our endeavors, regardless of whether we are dedicated to science or other equally worthy pursuits. We will consider that conclusion at the end of the book.

CHAPTER 1
"IGNORANT" GENIUS
Double Helix

[On the human genome] A more important set of instruction books will never be found.
James D. Watson[1]

The discovery of the double-helix structure of DNA, the substance of heredity, has been likened in significance to that of Darwin's theory of evolution. The principal actors in this discovery were James D. Watson (1928–) and Francis Crick (1916–2004). Watson wanted to make a discovery; Crick wanted to understand what life was. Watson was a genius in bringing together the biological interest in DNA and the possibilities of structural chemistry. It helped that he was somewhat ignorant of the limitations of recent techniques. Had he been fully versed in them, he might not have dared to initiate the determination of the DNA structure. In hindsight, Watson realized that one could not be fully qualified in an area before embarking on a project that would culminate in a big leap in science. Some of Watson's traits were especially helpful in his achieving success in this discovery and in the rest of his career.

The last sentence of Watson and Crick's seminal paper about the double helix has become a celebrated quotation in the scientific literature: "It has not escaped our notice that the specific

Figure 1.1. Double-helix sculpture by the Swedish sculptor Bror Marklund in front of the Biomedical Center of Uppsala University, Sweden. Photo by the author, 1996.

pairing we have postulated immediately suggests a possible copying mechanism for the genetic material."[2] Today, this copying mechanism is commonly known, whereas at that time, in 1953, the thought was revolutionary. The double-helix structure of deoxyribonucleic acid (DNA) came within a decade after the discovery that DNA was *the* genetic material.[3] When Oswald Avery and his two associates first pronounced in 1944 that the substance of heredity was DNA, few people noticed it and fewer yet were impacted by it. When, in 1952, the same pronouncement was made by Alfred Hershey and Martha Chase, and on the basis of less thorough experiments, it was enthusiastically accepted.[4]

Watson and Crick's paper in the April 1953 issue of *Nature* was barely longer than one page, and it stressed that its authors merely suggested—rather than determined—a structure. However, the structure had important novel features. One was that DNA consisted of two helical chains, each coiling around the same axis but

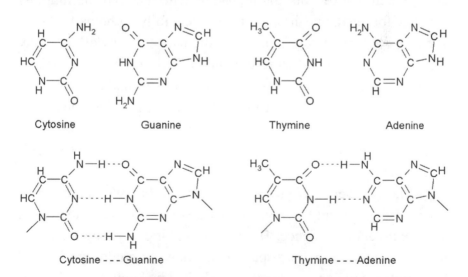

Figure 1.2. Top: The four nitrogen-bases in DNA. Bottom: Base pairs between pyrimidines and purines established by hydrogen bonding: cytosine–guanine and thymine–adenine.

having opposite directions and thus complementing each other. The other novel feature was the manner in which the two helices were held together through hydrogen bonds between the purine and pyrimidine bases. The bases were joined in pairs, a single base from one helix paired with a single base from the other helix. The two bases in a pair lay side by side, and the complementary pair of a purine base was always a pyrimidine and vice versa.

The report in *Nature* was illustrated by a majestically simple sketch.

The structure was consistent with all the information available by then: X-ray crystallography, model building, and chemical analysis of DNA. The latter indicated equal amounts of the purine and pyrimidine bases, regardless of the organism for whose DNA the composition was determined.[5]

The discovery of the double helix opened a new era in science with a direct route to the Human Genome Project four decades later, which mapped out the human genetic material. Its limitless benefits for medicine have not yet been fully fathomed. But in 1953, it was only a suggestion, and the painstaking work of many scientists was needed before the double helix became a certainty. For years, Watson had doubts about the structure, and it wasn't until the early 1970s that he had his first good night's sleep about the double helix. This was when he learned that reliable crystal structure determinations of DNA, finally, confirmed his and Crick's original suggestion.[6]

Although there were uncertainties about the 1953 discovery, it catapulted the twenty-five-year-old Watson to a place at the top of twentieth-century science. He was an ambitious young man who himself wondered in retrospect about how he could "go beyond [his] ability and come out on top."[7] He had had doubts all along as to whether he was bright enough, whether he would be able to solve a problem, and whether he would ever have original ideas. He was much sooner a genius than a great scientist. What happened to him was the fortunate confluence of many factors of being at the right place at the right time and, above all, being the right person for his self-ordained task. It certainly was not sheer luck, because it was his decision about what to do and where to continue his career when he faced competing options. But circumstances, too, favored what he decided doing. Peter Medawar, the Nobel laureate immunologist, remarked, "Lucky or not, Watson was a highly privileged young man."[8] It was less through his background at home than the environments he eventually found himself in and utilized that made him privileged.

Watson came from a nonpracticing Christian family with mostly Irish and Scottish roots. His family lived on the south side of Chicago in a neighborhood that was not very well-to-do, but it was not impoverished either. His parents were determined to help their two children get a good education. Watson, in his blunt style, referred to this as having been brought up in a "quasi-Jewish" atmosphere in which books were more important than material goods.

He went to schools that were not especially remarkable, and he breezed through them at an accelerated pace. He was not a child prodigy, but he participated in quiz competitions with remarkable success. He left his mother's Catholic faith by the age of twelve and graduated from high school at the age of fifteen. He became a student of the University of Chicago under its maverick president, Robert Hutchins, who did not care much for specialized instruction and placed the Great Books as the focus of college instruction. This broad-based education proved beneficial to Watson. Already in his youth, he was more ambitious than most other students, and when he found a subject that interested him, he was keener to learn more about it than anybody else. He did not mind when he saw that others might be more talented than he was; on the contrary, he sought out their company. If he could learn from others, he did, and he did not find it beneath him to imitate others if he found them worthy of imitation.

It was during his undergraduate studies that Watson read Erwin Schrödinger's influential book *What Is Life?*[9] It, more than anything, contributed to his transformation from a bird watcher/zoologist into a geneticist, which he remained for the rest of his life. By the age of nineteen, he had completed his college education and was considering graduate schools. The big-name schools were not kind to him, and perhaps there was nothing remarkable about him, except his eagerness, which may not have come through in his written application. He ended up at Indiana University in Bloomington in 1947, and as it turned out, Indiana was probably the best place for his further development. For a brief period, Bloomington offered top graduate education in modern biology. It had one Nobel laureate, Hermann J. Muller, and two future Nobel laureates—three, including Watson—in the same department at the same time.

Muller, who had been a professor at Indiana since 1945, was awarded the Nobel Prize for his studies of mutations by X-ray irradiation in 1946. His presence was a significant addition to the status of Indiana, but it impacted Watson less than the presence of Salvador Luria and others. Luria became Watson's doctoral mentor and would later win a Nobel Prize himself. Luria was a cofounder of Max Delbrück's school of phage study. The phage, short for *bacteriophage*, is the virus that attacks bacteria, and it was thought to be the best target for genetic research. Through Luria, Watson came under Delbrück's influence. He was not the only one, as Delbrück was a major influence on twentieth-century biology—not so much through his research or his ideas, which in most cases did not prove to be correct, but through his ability to challenge and inspire others. Schrödinger's *What Is Life?* was to a great extent an expression of Delbrück's views; Delbrück had transformed himself from a physicist into a biologist.

Despite being in the American Midwest, Indiana University provided Watson a diverse international environment with a strong European flavor. Muller was American with experience in Soviet Russia; Luria was a Jewish-Italian refugee who had escaped from fascist Italy; and another fellow student and future Nobel laureate, Renato Dulbecco, was a postwar immigrant from Italy.

Watson received a PhD degree at the age of twenty-two. His dissertation examined whether phages that had been inactivated by X-rays could be reactivated. His work was unremarkable, which was—paradoxically—a blessing in disguise, because he did not feel any pressure to continue his doctoral project. Nor did he feel pressure from others to achieve anything extraordinary—yet. This was a period for absorbing knowledge and information, which he did mainly from personal encounters. He had no inhibitions about walking up to anybody, even if that person was the greatest name in the field, if that person possessed the information he sought.

Upon earning his doctorate, Watson left for Denmark for post-doctoral studies. He was not happy with his first assignment, so he moved to another laboratory, but the project there did not satisfy him either. At that point, in the spring of 1951, he attended a meeting in Naples where he listened to Maurice Wilkins talk about the X-ray work on DNA at King's College in London. Watson glimpsed Wilkins's photograph of an X-ray diffraction pattern and decided "to move to an X-ray crystallographic lab devoted to macromolecules."[10]

Watson's decision to switch to working on the structure of DNA was significant for at least two reasons. One was mundane: though the funding agency for his postdoctoral fellowship opposed his move, Watson disregarded its opposition and subsequently lost his fellowship. The other reason was that at this point he hardly knew anything about X-ray crystallography, let alone its application to biological macromolecules. This was the time when the giant of science,

Figure 1.3. James D. Watson in June 1953 at Cold Spring Harbor,
New York, shortly after the discovery of the double-helix structure
of DNA. The model of the double helix is in his left hand.
Photo by and courtesy of Karl Maramorosch, Scarsdale, New York.

Linus Pauling, was struggling with his protein structure, ultimately leading to his discovery of alpha helix. This was also the time when the star-studded British team of W. Lawrence Bragg, Max Perutz, and John Kendrew had already published a plethora of erroneous models for protein structures (see chapter 4). Watson's ignorance must have contributed to his bold decision for his next career move, but it was also a sign of genius that he embarked on this route.

The use of the term *ignorance* here calls for a caveat. Watson was, of course, not ignorant in many aspects of the research skills needed to discover the structure of DNA. He was not ignorant in recognizing the importance of DNA and its structure. But he was not clear about the possibilities and limitations of structural chemistry and X-ray crystallography at the time. Even the experts found it very difficult to elucidate the structure of the supposedly easier target proteins (as is illustrated in chapter 4). There is nothing wrong in being ignorant in some aspects of the research one is about to conduct: if we were all well-versed in every aspect of our scientific research, it could hardly be called scientific research. Rita Levi-Montalcini might have had Jim Watson in mind and his initiating the work on the DNA structure when she stressed in her autobiography the importance of underestimating the "difficulties, which cause one to tackle problems that other, more critical and acute persons instead opt to avoid."[11]

Once Watson had decided on his project, he had to choose the venue for it. There were not many places at the time where he could pursue his quest for the DNA structure. Pauling's laboratory at the California Institute of Technology (Caltech) was a possibility, and there were the British laboratories—King's College in London and the Cavendish Laboratory in Cambridge. Watson wanted to remain in Europe, and the next choice, between London and Cambridge, was not difficult, as both tradition and the name of W. Lawrence Bragg favored Cambridge. Wilkins, at King's College, may have lit the spark, but he did not attract Watson to join him. Looking back, the partnership with Francis Crick proved unsurpassable, but Watson could not have known that at the time.

The change from being on the periphery of science in Denmark (periphery, that is, in molecular biology, not, of course, in Niels Bohr's physics) to a world-class center in Cambridge was to Watson's liking. No sooner had he arrived than he teamed up with Francis Crick, his assigned roommate. Crick had a background in physics, was full of ideas, and had been engaged half-heartedly in an unexciting project. He was soon infected by Watson's vision, and they formed one of the most remarkable partnerships in science history.

It fell on Watson—as a proxy—to present the results from the experiments of Hershey and Chase in 1952, which reinforced Avery and his colleagues' findings that DNA was the substance of heredity. Also, in 1952, the biochemist Erwin Chargaff visited the Cavendish Laboratory and told Watson and Crick about his seminal experiments. Chargaff's discovery, which had direct relevance to Watson and Crick, was that the DNA bases adenine (a purine) and thymine (a pyrimidine) occurred in roughly equal amounts, and so did the bases guanine (a purine) and cytosine (a pyrimidine), regardless from which organism they had been extracted.

Scientists congregated in Cambridge and were anxious to share their latest findings with the researchers there, as if seeking their approval. Then yet another fortunate circumstance occurred: Linus Pauling had sent his son Peter to Cambridge, and Peter became friendly with Watson and Crick. The young Pauling was happy to carry the news from his father about progress at Caltech to his new friends. Also, Watson and Crick welcomed a new roommate at the Cavendish in the person of American chemist Jerry Donohue, who put them on the right track about the preferred chemical forms of the bases in DNA.

Watson knew hardly any chemistry at the time of the double-

Figure 1.4. Francis Crick in the Cricks' home in La Jolla, California, in February 2004. Photo by the author.

helix discovery, but he was always ready to learn what he needed to know. Much later, when he won the Nobel Prize and had to give a Nobel lecture, he chose an expressly chemical topic for his presentation, as if correcting the popular image about his ignorance of chemistry. Incidentally, Crick's Nobel lecture was not about the discovery of the double helix; it was about the genetic code. Only Wilkins spoke about the discovery of the double helix, in which he had the most minor role among the three in 1953.

Another scientist, Rosalind Franklin, had been more directly involved, but she had died by the time the Nobel award was granted for the double helix. Had she lived, she might or might not have been included. There is a three-person limit of awardees in any given category of the Nobel Prize, and in the early 1960s, Franklin's contribution had not been recognized to the extent it has since. Her negative portrayal in Watson's book *The Double Helix*[12] a few years later would generate a large amount of research into her contributions, which has been followed by widespread appreciation for her work.[13] She was a crystallographer at King's College, and a strong animosity developed between her and Wilkins. She and her student Raymond Gosling produced superb X-ray diffraction patterns of DNA samples. Wilkins showed the best of her diffraction plates to Watson without Franklin ever learning about this act of betrayal. Then, Max Perutz, while serving on a review committee evaluating the work at King's, informed Watson and Crick about the progress of Franklin's work, again without her knowledge. She was thirty-eight years old when she died of cancer in 1958. After Franklin's death, her former associate and future Nobel laureate Aaron Klug examined her lab journal and discovered that she was much closer to solving the DNA structure than had been believed.[14] Her forced departure from King's College, however, terminated her work on the DNA structure before she could have completed it.

Watson was very lucky, but he worked hard at finding his luck.

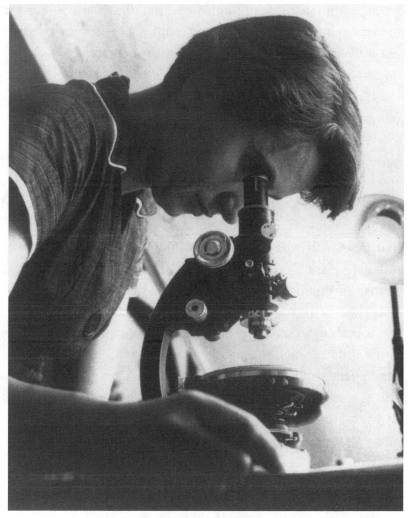

Figure 1.5. Rosalind Franklin. Photo by Henry Grant, AIIF.
Courtesy of Aaron Klug, Cambridge, United Kingdom.

He always had the right mentors, supporters, and partners. He ultimately found the right wife, the right venues for remaking a
research place in his own image, and most of all, the right shoulders to stand on in order to look farther. Those shoulders belonged
to Delbrück, Pauling, and others, including, unwittingly, Franklin.
There has been much effort to demonstrate that there was nothing

wrong with having communicated Franklin's data by third parties to Watson and Crick, though it was at least questionable whether or not it was "legal."[15] Whether or not it was "moral" was a different question, and it has never been suggested that it was.

It is revealing how, decades later, Watson referred to the episode in which Wilkins had showed him Franklin's crucial experimental evidence. Watson had gone to King's College to spread the news about Pauling's erroneous triple-helix model of DNA. On that occasion, in Watson's words, "Maurice [Wilkins]—bristling with anger at having been shackled now for almost two years by Rosalind's intransigence—let lose the heretofore closely guarded King's secret that DNA existed in a paracrystalline (B) form as well as a crystalline (A) form."[16] So it was not so much that Wilkins altruistically shared information with Watson for the sake of advancement of science; rather, his motives were based on anger and revenge.

In addition to Franklin and Gosling's X-ray patterns, Watson and Crick utilized Pauling's approach of relying on all available structural chemistry in their quest for the DNA structure. This was perfectly normal and constituted a brilliant example of how the next discovery builds on previous discoveries, utilizing published data and techniques. Watson and Crick needed "only" to put together all the relevant information after they had performed the most crucial act of posing the right questions. Theirs was an unusual but very efficient approach to research at that time: using other people's measurements, techniques, experimental results, and conclusions. Science works this way, except it usually goes more slowly and less obviously, but as Isaac Newton explained, he saw farther than his predecessors because he stood on the shoulders of others.[17]

Watson and Crick's working style appeared unorthodox to many. They seemed sloppy, did not seem to be working too hard, and appeared as if they had plenty of free time for entertainment. At times they behaved as if they were underemployed—not the

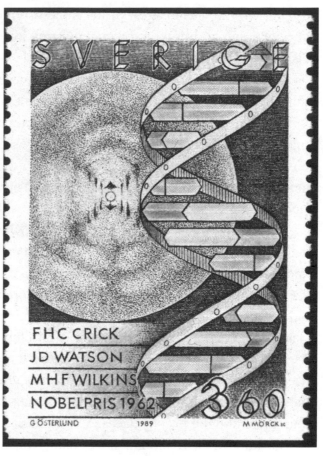

Figure 1.6. Swedish stamp commemorating the 1962 Nobel Prize for Francis Crick, James D. Watson, and Maurice Wilkins with Watson and Crick's double helix on the background of Rosalind Franklin and Raymond Gosling's X-ray photograph of DNA.

usual image of the mad scientist who lives for his work day and night. Furthermore, they seemed too interested in scientific gossip and not enough in learning from the scientific literature. However, there is no definition of what constitutes the most efficient approach to research. And the unconventional features of Watson and Crick's approach turned out to be an excellent way to attack the problem they were working on.

We have already met Max Delbrück, the physicist turned biologist and one of Watson's idols. Delbrück formulated the Principle of Limited Sloppiness: If one is very sloppy, that is bad, but striving for too much rigor might hinder advances.[18] Crick formulated his idea about the advantages of listening to gossip because the grapevine might bring in crucial information that had not yet reached the degree of finality that would be fit for publishing. Finally, hard work and hard thinking do not necessarily appear the same on the surface, while the latter is no less necessary than carrying out yet another experiment or computation. Not every environment would have tolerated Watson and Crick's way of doing science as well as the Cavendish Laboratory did. As Max Perutz, their nominal boss at the Cavendish, put it, "There is more than one way of doing good science."[19]

Watson and Crick never explicitly acknowledged that Watson had had access to Franklin's data, not even in their groundbreaking April 1953 *Nature* paper, which was as much a breach of ethics as snatching the information itself. Watson ignored—whether knowingly or just because he did not care—many minor and not so minor societal conventions. Some of this was on purpose. Legend has it that he was so absentminded that he often forgot to tie his shoelaces. But, according to gossip, it had been observed on occasion that Watson, when arriving at a party and just before entering the house, would untie his shoelaces.

His idiosyncrasies might have made Watson unwanted company, but the opposite happened: they enhanced his popularity. So did many of his mannerisms that went against accepted norms. He mumbled in his lectures, often speaking to the blackboard rather than to the audience, and in a voice hardly audible; yet his talks were eagerly awaited and attended by throngs. He was a poor dresser but was invariably invited to gatherings. He was clumsy and awkward with girls, but the Cambridge ladies threw themselves into helping him find dates. He seemed to have the ability of turning his private problems into subjects of community concern.

Watson thought a lot about how to succeed in science. He wanted success, and he thought about the Nobel Prize. His is one of the rare cases when such an overt craving for fame and recognition proved fruitful rather than counterproductive. Fame was a driving force for him, and he set up rules that ensured success. And it was not at all theoretical, because he practiced his rules. Watson summarized his prescriptions in over a hundred rules in his book *Avoid Boring People*, and that title epitomized one of his favorite rules.[20] For scientific success, he had set up six rules that he pronounced long before his rule making had become extensive:

1. Avoid dumb people.
2. Be prepared to get into deep trouble.
3. Be sure you always have someone up your sleeve who will save you when you find yourself in deep s–.
4. Never do anything that bores you.
5. Expose your ideas to informed criticism.
6. If you can't stand to be with your real peers, get out of science.[21]

When I lecture about Watson in my course on the great discoveries in the twentieth century, I tell my students what would happen if Watson opened the door to our auditorium and looked for a place to sit down: He would look around searchingly and would sit next to the person in the audience whom he perceived to be the most interesting. At this point in my lecture, there is usually a little commotion; my students look around as if assessing themselves and their peers, and sometimes one of them shifts in his seat as if making room available for Watson to sit next to him (it's invariably "him" rather than "her"). The saying is attributed to Watson that if you are the smartest man in the room, you are not in the right room—Watson's lack of vanity made him always seek out the smartest company to be in.

Watson's keys to success are to be found, however, in a broader domain of traits than the six points listed above. He had the ability to distinguish between the important and the unimportant, and he always found time for relaxation. These two qualities may not be in contradiction. Watson economized his time, but when he was doing something that he judged to be truly necessary, he spent his time on it liberally. Thus he was very patient when he was cutting out paper as the bases for his model when he was on the verge of the discovery of base pairing in DNA. He paid meticulous attention to the minutest details when writing his textbooks. He devoted a lot of time to the back-and-forth exchanges with his colleagues and friends as he was preparing the publication of his book *The Double Helix*. And he paid the most careful attention to the smallest details in the planning of new constructions and renovating old buildings at the Cold Spring Harbor Laboratory (CSHL) he eventually headed.

Figure 1.7. James D. Watson (right) and the author in the Manhattan home of Watson and his wife, Liz, 2010. Photo by and courtesy of Magdolna Hargittai.

What Watson did *not* do is also noteworthy. There are scientists who, once they find a fertile area of research, exploit it to the fullest; once they establish a new methodology, they apply it to whatever is appropriate. For Watson, it was never a problem to withdraw from an area of research when his work was becoming repetitious. After the discovery of the double helix, and only after having made sure that everybody saw its biological implications, he moved on. His experience with the study of the structure of RNA and with the quest for the messenger RNA made him realize that instead of trying to top his previous feat, he should be seeking his success elsewhere.

A career that many others would have cherished, including a Harvard professorship, could not have satisfied him in light of the double-helix triumph, so Watson proceeded in two other directions: authoring books and becoming a science administrator. In both, he became immensely successful. His textbooks covered new ground and were innovative not only for their content but also for their style. His narration of the double-helix discovery showed the process of scientific research in a way that nobody before him had been capable of showing or had dared to try. Cold Spring Harbor Laboratory (CSHL) has become highly respected, as has its Watson Graduate School, and it has also become Watson's shrine.

CSHL has carried out research in cancer, in neurological and other major diseases, in food resources, and in efficient biofuels. Its past successes include Barbara McClintock's Nobel Prize–winning discovery of mobile genetic elements and Richard J. Roberts's Nobel Prize–winning discovery of split genes (this was a shared award with Philip A. Sharp). However, whether CSHL becomes a lasting success will be decided only when Watson is no longer around to be its supreme leader, regardless of his actual title. He had built up a lot of hostility there because of his methods of promoting successes through competition between members of the same group and between groups of the same laboratory. Nothing seemed too sacred to him in pursuit of success.

Watson was seldom a player in politics at the national level. He preferred remaining in the realm of science, though there were a few exceptions. When President Richard Nixon declared his War against Cancer, Watson pointed out the futility of the project. He showed that the money could not be spent wisely if basic understanding was lacking as to the causes of the different cases and the mechanisms of action. He acquired a yet more prominent role in the Human Genome Project between 1990 and 1992—a brief period of time but a crucial one, since this was the start of the project. Otherwise he was hardly involved in politics in any visible way, and his additional public appearances made headlines for some shocking but inconsequential statements, like the one that fat women have better sex lives than others. Mostly, though, he was restrained and knew what he could say publicly and where to draw the line to keep his views private, with due consideration for his fundraising role at the Cold Spring Harbor Laboratory.

It appeared a little out of line—and somewhat off topic—when Watson reflected on the unfortunate story of Lawrence Summers's statement about women in science in his new book *Avoid Boring People*. While president of Harvard University, Summers in 2005 suggested that the underrepresentation of women in the top levels of academia was due to a "different availability of aptitude at the high end." It was a meaningless and offensive generalization for which Summers then apologized, and which contributed to his eventual resignation from his position. Watson's reflection on the story was that in a more normal world there should be a lot of research put into examining questions like this, rather than brushing them off in their entirety.

Soon a story involving Watson came out that had an air of déjà vu. Prior to a tour in England in October 2007 to promote his book,

Watson spent six hours with a journalist who came to interview him at CSHL. They chatted and exchanged travel experiences, and during their whole encounter, the journalist's tape recorder—unknown to Watson—was running. In her account in the *Sunday Times*, she quoted some shocking remarks by Watson. He told her that he was "inherently gloomy about the prospect of Africa," noting that "all our social policies are based on the fact that their intelligence is the same as ours—whereas all the testing says not really."[22] There were some other, similar remarks, and after the ensuing outrage, Watson profusely apologized and resigned his chancellorship at CSHL. The CSHL leadership sharply distanced the institution from him.

The eighty-year-old scientist publicly expressed his gratitude to CSHL for letting him stay in his home on CSHL grounds, though he was otherwise humiliated. He could no longer make statements to the media on his own, and so a publicist was hired to handle his interviews. Watson invited me to a rare interview he gave to a journalist in March 2008. The publicist was on hand to control not only Watson but also the journalist, who was obviously ignorant about Watson as a scientist. It was an embarrassing situation, especially when compared with Watson's prior independence of thought and style.

Due to Watson's age, this event could have signified the closing of his career and would have made an especially sad ending. Watson, however, was not done yet: he persevered. He managed a gradual comeback, and the former whiz kid, now an octogenarian, has lately been active again, traveling, giving talks, and raising funds—for CSHL. James D. Watson is still going strong. The quest for success has been his driving force, and it has kept him going.

Figure 2.1. Gertrude B. Elion in her office in
Research Triangle Park, North Carolina,
1996. The picture on the wall at the top is a
reproduction of the drawing by the British
caricaturist James Gillray, titled *The Gout*.
Photo by the author.

CHAPTER 2
PUSHED BY PERSONAL TRAGEDY
Lifesavers

**. . . my daughter, Tiffany is alive and well today
because of you and your research.**
From a letter to Gertrude B. Elion[1]

*Gertrude B. Elion (1918–1999) decided at first to become a
chemist and a scientist at fifteen years of age when she witnessed
her grandfather dying from cancer and later when she lost her
fiancé to an illness for which there was no cure at the time. She
vowed that she would devote her life to finding new medications
to save people. She encountered obstacles on her way to achieve
her goals: her family was so poor that she could not afford to
complete her doctoral studies, and she was a woman, so it was
hard to find employment that would allow her to grow. Nonetheless, she overcame all the hurdles and not only created new lifesaving drugs but, along with her mentor George Hitchings,
worked out a whole new approach to drug design that would help
other researchers in their quests for new drugs. When Hitchings
and Elion were awarded the Nobel Prize in Physiology or Medicine in 1988, Elion suddenly found herself in the limelight.
People who had been using her drugs learned about her as a
person, and came to see those medications as an offering by a
human being. Letters of gratitude poured in for Elion. She
became a hero.*

On the same day I heard the news about Gertrude B. Elion's death over the radio in Wilmington, North Carolina, I received her last letter. It was on Monday, February 22, 1999. My wife, Magdi, and I had visited her three years before in Research Triangle Park, North Carolina, where we recorded a conversation with her, and we maintained a correspondence. At one point I asked her for a few photos of George Hitchings, her former mentor, about whom she spoke with great respect and who had recently died in 1998.

In the late fall of 1998, Elion still had a busy schedule: she attended meetings, gave talks, and had to squeeze in a cataract surgery among her many obligations. In early February of 1999, she wrote me that she had had a successful surgery and a pleasant trip to Europe behind her and was ready for the Hitchings project. She sent me pictures on February 17 and some text on February 19, but then, on February 21, she passed away.

I was a visiting professor at the time at the University of North Carolina at Wilmington, and on that Monday I had an evening general chemistry class. During the day, the news about her made me reflect on our conversations, and I was still thinking about her as I entered the auditorium. I began my lecture by telling the class about Elion. I thought there was plenty of relevance: she was a Nobel laureate chemist who lived and worked during the last decades of her life in North Carolina. I told my students about the drugs she discovered and showed her picture. I did not speak for more than a couple of minutes when I sensed some impatience, and I was almost done when a student interrupted me, "Aren't we going to review the test problems?" I always encouraged my students to interrupt me with their questions, and the interruption did not upset me. Also, I knew Elion was a patient and tolerant person. My class was an evening class, and most of my students came after their day's work, tired. Elion would have especially sympathized with them, because she had had a difficult road in obtaining her education and often had to study while holding a job.

At the time of her death, Elion was scientist emeritus and consultant at Glaxo Wellcome, Inc., in Research Triangle Park. She took this position after she retired in 1983 from her job as head of the Department of Experimental Therapy at the same company (at that time it was Burroughs Wellcome). She enjoyed an exceptionally rich career in discovering new drugs, but she almost did not take this path. It took a great effort for her to get through high school and college. The difficulties facing a poor Jewish girl of her generation embarking on a research career can be illustrated by the fact that, in spite of all her hard work, she could not advance beyond a master's degree. As it was, she certainly defied what could have been prescribed for her when she was just starting out.

Elion was born in 1918 in New York City. She and her brother Herbert, six years her junior, went to public schools at the time when the New York public school system was famous for its scholarship. It has been noted that "twenty-four students born of poor, immigrant Jewish families [who] graduated from New York City public high schools in the second quarter of this century [became] winners of the Nobel Prize, a record. . . ."[2] Elion went to a girls' high school, jumped a couple of grades, and entered Hunter College, a women's school at the time, in 1933. She decided to major in chemistry.

She was a voracious reader, and her interest in science came mostly from her books. Becoming acquainted with great scientists from her readings, she considered Louis Pasteur and Marie Curie to be her idols. Paul de Kruif's book *Microbe Hunters* had a profound impact on her. The book was published in 1926, became an instant bestseller, and remains in print to this day. It is about natural scientists who dedicated their lives to uncovering nature's secrets. It is written in a romantic style, perhaps less appealing to the present generation of young people than that of Elion's time, but it is still a captivating read. Among the famous scientists of Elion's generation, this book was a major influence on gifted children in choosing a career in science.

Elion, however, had an additional, specific reason for her resolution to become a scientist. She was very affected by her grandfather's death. He died of stomach cancer, so she vowed to devote her life to finding cures for people who were suffering from cancer and other diseases. During Elion's youth, it was not very easy to get into Hunter College; in addition to the requirement of New York City residency, the applicants were supposed to have at least an 85 percent average in high school; however, there was no tuition. Analogously, the boys had City College. Today, circumstances are different; both City College and Hunter have become coeducational; there is open enrollment and tuition is charged.

Elion's parents had both descended from scholarly families and had both arrived in the United States in their early teens; he—Robert Elion—from present-day Lithuania, and she—Bertha Cohen—from present-day Poland. Robert had studied dentistry in New York and was good at investing, so they did well until share prices collapsed in the fall of 1929 at the New York Stock Exchange, and they lost everything. The Great Depression hit the family, and hardship followed, but they would never have considered their children not getting a college education. They had traditional respect for education and knew it was the key to success. Elion always regretted that her parents died before they could witness her successes.

When Elion graduated from Hunter—and she did so with the highest distinction—she applied to fifteen graduate schools in the hope of winning an assistantship or fellowship, but nothing came of her efforts. It was a difficult time because of the Depression, but sex discrimination and anti-Semitism also played a part. Elion found some teaching jobs and even worked for free as a laboratory assistant to a chemist in order to gain some practice. Eventually she became a graduate student at New York University, paying her way by working as a teacher and doing her thesis research on nights and weekends. She received her master of science degree in 1941. By then, World War II was raging, and the United States—after an initial period of neutrality—soon became a combatant. Employment for female chemistry graduates had become easier, so Elion again began looking for a job. When she was interviewing at a prospective company, her interviewer told her, "Oh, you only have a Master's degree, therefore, we should tell you now, this [position] is as high as you can go."[3] The encounter made Elion think that they had no idea how good she might be or if she was any good at all, or that she did not even think about how far she could go. Though she was offered the job, she did not take it.

It was after she had graduated from New York University that she met her future fiancé, Leonard Canter, who had graduated from City College and won a fellowship to study abroad. They grew close, and when he returned from his trip, they decided to get married. Tragedy then struck: he became ill with a condition that was severe at the time but that might have been prevented with antibiotics developed a few years later. He had acute bacterial endocarditis, an infection of the heart, and he could not be saved. His death devastated Elion, and she never married. This was not because of any vow she had taken but because she never found anybody who could have been compared with her lost fiancé. But the tragedy strengthened her resolve and dedication in finding cures for sick people.

Her interview in 1944 with George Hitchings at Burroughs Wellcome was entirely different from her previous experience. Hitchings told her about his work and made it sound exciting. He offered her a job, and she went to work for him. Hitchings kept his word and let Elion, who had an insatiable thirst for knowledge, learn everything she wanted to learn. He never told her, "It is not your business to do this or that," and he never said, "You are not a pharmacologist, not a virologist, not an immunologist," so Elion became all of them.[4]

The company itself was unusual. Two American-British pharmacist-entrepreneurs, Silas M. Burroughs and Henry S. Wellcome had founded it in England in 1880, and the company earnings were to benefit the Wellcome Trust to support medical research and related activities.[5] Hitchings was a perfect scientist for the company because he was dedicated both to scientific research and to discovering drugs, and Elion became his perfect partner in this endeavor.

Hitchings quoted Wellcome in his speech at the Nobel Banquet in 1988, "If you have an idea, I will give you the freedom to develop it."[6] Hitchings earned his PhD degree at Harvard University in 1933, and his doctoral work was in the area of nucleic acids. These substances were considered rather dull at the time but became a focal point in biochemistry in the 1940s, so Hitchings's work proved to be a forward-looking project. After a series of unremarkable jobs, Hitchings joined the Wellcome Research Laboratories in Tuckahoe, New York, in 1942. He was appointed head of the Biochemistry Department, and for some time he was the only member of his department.

The year Elion joined Hitchings and his first assistant, Elvira Falco, was the same year that Oswald Avery and his two associates at Rockefeller Institute (as it was then called) published their seminal paper asserting that deoxyribonucleic acid (DNA) was the substance of heredity. Initially very few researchers were interested in this finding; the notion that proteins carry the genetic information from generation to generation would dominate the thinking of biologists for a few more years. Besides, the double-helix structure of DNA would not be discovered by Francis Crick and James

Watson until 1953. But Hitchings was among the few who recognized early on that consideration of nucleic acids might be fruitful for drug discovery.

Hitchings broke away from the traditional drug research of trial and error in which selected compounds that had been identified as having biological activity were modified until derivatives that had the desired effects could be obtained. Hitchings had a new approach, which was fully justified in being called *rational drug design*. His idea was to interfere with the existence of bacteria, tumor cells, and malarial parasites by targeting their DNA with substances that were similar to their building blocks. Such building blocks would be built into the DNA of the organisms that he wanted to prevent from multiplying. It was not only a very ambitious approach; it was also a very modern one, because it made use of the latest information about the role of DNA.

The nucleic acids, both DNA and the ribonucleic acid RNA, consist of nucleotides that are built from a phosphoric acid, sugar, and a nitrogen-containing base. The phosphoric acid and the sugar are the same in all nucleotides; their differences occur in the bases. There are five different bases; four each for DNA and RNA; and only one of the four bases differs in the two nucleic acids. The bases are of two different chemical classes, purines and pyrimidines. For DNA, the purines are adenine and guanine, and the pyrimidines are cytosine and thymine. (Their chemical formulas were shown in the preceding chapter.)

In order to attempt to fool the DNA of bacteria, tumor cells, or parasites, new compounds had to be created that resembled the bases in the DNA of these organisms. These compounds had to fulfill two requirements: One was that they should be similar to the bases of the DNA of the target organisms so that they could incorporate them. The other was that these "misleading" building units would hinder the proper functioning of their host DNA but would not be toxic for the humans—whom the drugs were supposed to protect; that is, to cure, in the first place.

No wonder young Elion was fascinated by such a project and threw herself into the first task Hitchings gave her: the preparation of various purine derivatives. In synthetic organic chemistry, the German literature was very strong, and Elion made good use of her knowledge of German. She used to speak Yiddish with her grandfather and built on this knowledge at college when she took up German. She became proficient in preparing the substances Hitchings had designed, and she participated increasingly in scheduling their next tasks.

She did not stop at preparing new compounds; she also wanted to test her compounds for biological activity. Again, it was very fortunate that Hitchings never stopped her from broadening the scope of her work. Theirs was an ideal cooperative relationship, and for the first decades of their joint work it was impossible to separate their contributions. Hitchings never praised her work, but at every pertinent occasion he helped her get promoted. It became almost a routine: when he was being promoted, she followed him in the position he had just vacated. Hitchings was never disappointed, and neither were their superiors, because Elion performed impeccably in her rising responsibilities. Nonetheless, she never stopped worrying about her missing doctorate.

At some point, she decided to make another attempt at pursuing her doctoral studies and signed up at Brooklyn Polytechnic Institute. She was studying part-time and thus needed to commute between her three locations: her home with her family in the Bronx, her workplace in Westchester County, and her school in Brooklyn. Her regime was working out well, but then the dean of her school informed her that she could continue her studies only as a full-time student and would have to give up her daytime job. This was something she could not and would not do. It was painful, but she made her decision: she gave up her dream of getting a doctorate. Her only consolation was that the other dream, even more important, remained: her quest to discover new drugs.

In time, though, she did become a doctor by receiving many honorary doctorates. The first such honor was bestowed on her in 1969 by George Washington University, where scientists did considerable research in an area very much related to Elion's work. Hitchings accompanied her to the ceremonies where she'd be receiving her certificate. In time she would receive many other honorary doctorates, including one—what an ironic twist—from Brooklyn Polytechnic Institute.

Success did not come easy though, and it did not come fast. Elion's first project was the preparation of drugs that would tease the formation of bacterial DNA. Looking back, they used a rudimentary approach, which was both time-consuming and labor-consuming, but, eventually, it worked. When the scientists had the compounds they thought might work, they tested them. For the antibacterial tests, they used the bacterium *Lactobacillus casei*. First, they gave these bacteria the natural nutrients containing the building blocks of their DNA, and the bacteria grew and multiplied. Then they fed these bacteria with modified substances that they had prepared. The bacteria would consume them, build them into their DNA, and then their functioning would stop. In other words, Hitchings and his team succeeded in deceiving the bacteria.

When Hitchings and his two associates had achieved successful candidate compounds for a particular purpose, as antibacterial agents, they went on testing whether the compounds would work on other systems as well, such as animal tumors. They had an agreement with the Sloan-Kettering Institute, where scientists had animal tumors growing in mice; the new compounds were tested at the institute. The mouse experiments were useful for an additional reason: they showed whether or not the compounds were toxic. The fortunate cases were when the compounds affected the tumor but not the mouse. Further testing was possible for malaria at the Wellcome Research Laboratories in London.

The first breakthrough arrived in 1950. Following successful animal experiments, a purine derivative (in this case, a modified guanine molecule) appeared to be working for acutely ill human leukemia patients at the Sloan-Kettering Memorial Hospital. The fate of one of the patients was especially apparent. She responded well to treatment, went into remission, and appeared to be cured. At this point, her treatment was stopped. After two years of remission, however, she relapsed and died. Scientists eventually understood that the treatment should not have stopped during remission, but by then, it was too late. There was a problem with the particular purine derivative in that it had strong toxicity and caused severe vomiting in the patients who took it. Elion continued her search and produced and tested over a hundred purine derivatives.

The first big success came with a relatively simple modification of the purine base guanine. When Elion replaced the oxygen in this molecule with sulfur and removed the amino group (NH_2), she produced what was then called 6-mercaptopurine, abbreviated as 6-MP. This caused remission in children with leukemia, although a relapse followed in every case. The problem, then, was that the remission proved temporary; the children soon relapsed and died. It

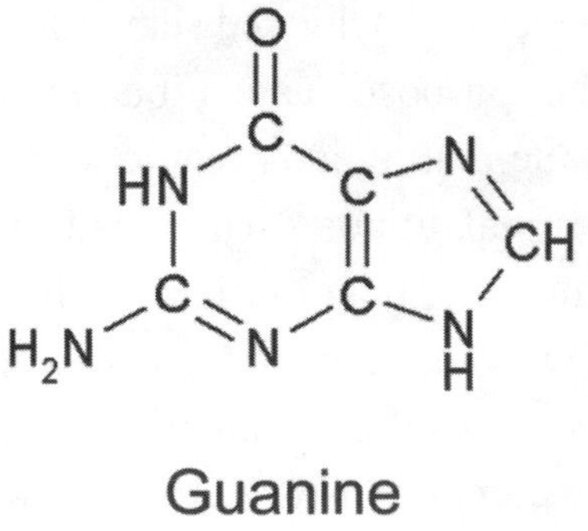

Guanine

Figure 2.2. Structural formula of guanine not yet modified.

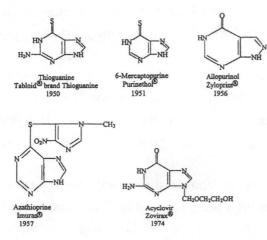

Figure 2.3. Structural formulae of five drugs figuring in the text, drawn in accordance with Gertrude Elion's sketches.

was a trying experience for Elion, yet the temporary remission gave her hope because it indicated that she and her team were on the right track. The progress was undeniable, and the Food and Drug Administration (FDA) approved the drug for commercial purposes in record time.

At this point, Elion and her team could have continued the search for improved drugs by blindly changing the chemical composition of 6-MP. Instead, they did something different—an approach that would become characteristic of their rational drug design. Elion suggested to Hitchings that they investigate the metabolism of 6-MP in the human body. The goal was to understand what happens to this chemical within our bodies. They expected that this understanding would point the way to an improved version of the drug.

Ideally, this would have been the job for pharmacologists rather than for a chemist like Elion, but she did not want to disrupt the work of pharmacologists from another department just because she was "curious." And curious she was! Thus, instead of finding pharmacologists who would be willing and available to conduct the studies, Elion acquired the necessary knowledge and techniques to do the investigation herself. She prepared radioactive 6-MP so that its movement could be traced, put it in the mouse, collected the mouse's urine, and determined what happened to the drug. She synthesized the metabolites, that is, the products into which the drug had converted, and determined their composition. It took a long time to get to the bottom of this metabolism, because the technologies were much less sophisticated than they are today. But she finally understood the process and could continue her rational drug design with enhanced zeal.

To appreciate what these metabolite studies led to, we need to make a brief detour. At Elion's time, Burroughs Wellcome was an ideal place for the kind of research she and Hitchings conducted. The company had big moneymaking drugs, so they could encourage research on a broad scale. Even though nobody could have expected a lot of profit from drugs for treating childhood leukemia, the company encouraged Hitchings and his team in continuing their search in that direction. Eventually, however, some of the drugs from the Hitchings-Elion team did become highly profitable. One of them originated from the 6-MP metabolite studies, and it was to become the drug used for treating gout, which would turn into a major moneymaker for them.

One of the most conspicuous decorations on the wall in Elion's office was a reproduction of a drawing by the British caricaturist James Gillray titled *The Gout* (see fig. 2.1, p. 50). The artist himself suffered from this illness, and other gout victims stated that the drawing illustrated their pain in a most expressive way. The picture shows a vicious beast attacking a foot, ejecting arrows from among its teeth, while the beast's hands clamp the foot mercilessly. Gout causes a burning pain in the joints, most often in a big toe. It is a consequence of too much uric acid in the blood, which leads to the formation of uric acid crystals that accumulate in the joints. Uric acid is a by-product of purine metabolism; it builds up in the blood and is deposited in the tissues, causing painful inflammation.

So how did it happen that the 6-MP metabolite studies would lead to a cure for gout? As Elion and her team were trying to make the beneficial effects of 6-MP last longer, they found that the drug broke down within minutes from the time it was consumed, and the culprit in breaking it down was an enzyme called xanthine oxidase. Once this was understood, the idea of stopping xanthine oxidase followed; that is, the team had to inhibit this enzyme. The chemists soon came up with a compound called allopurinol, which inhibited the enzyme, thereby giving longer life to 6-MP. However, as cures with drugs are never simple, another problem developed. The longer 6-MP stayed in the organism, the longer its bad side effects kept occurring. Therefore, it was not a preferable approach to administer allopurinol for this purpose.

Yet when allopurinol was administered along with 6-MP to patients with a high level of uric acid in their blood and urine, the level of the uric acid content in their blood decreased. The possibility of reducing the production of uric acid in the organism was an important observation, and it led to an unexpected application: a new and effective treatment of gout, which the FDA approved in 1966.

In the meantime, Hitchings and Elion made another guanine derivative called thioguanine, in which only the oxygen of guanine is replaced by sulfur, and the amino group is not removed. They eventually realized that either 6-MP or thioguanine should be administered in combination with some other drugs, rather than alone. Thioguanine later found its most useful function in treating adults for another kind of cancer, myelocytic leukemia, a life-threatening disease in which cancerous cells replace normal cells in the bone marrow.

Curiously, the drugs introduced by the Hitchings-Elion team often found applications in solving more than one medical problem. This initially came to the team as a surprise, but it was a consequence of their rational approach to drug design and broad-scale testing. Since they were targeting the DNA of the culprits of various diseases and disorders, the cure they were offering may have been more comprehensive than originally envisioned.

Another research group in a Boston institution, for example, found 6-MP dramatically useful in muting the body's immune response following organ transplantation. This was first evidenced in a dog whose life was extended by treatment with 6-MP following kidney transplantation. Eventually, the drug Imuran, a derivative of 6-MP, became recognized as an immune-system suppressant. Imuran was mentioned with great appreciation in plastic surgeon Joseph E. Murray's Nobel lecture in 1990 about the first successful organ transplants in humans. Murray was awarded the Nobel Prize in Physiology or Medicine for his discoveries concerning organ transplantation. Before Hitchings and Elion produced Imuran, Murray could successfully transplant kidneys only for identical twins, without the transplant getting rejected. Imuran has made it

Figure 2.4. From left to right: the dogs Tweedledum and Tweedledee, unknown, Roy Calne, the dog Titus, Gertrude Elion, the dog Lollypop, George Hitchings, Donald Searle, E. B. Hager, and Joseph Murray. Courtesy of Gertrude B. Elion and Katherine T. Bendo Hitchings, New York City.

possible to perform this procedure for hundreds of thousands without such a limitation.[7]

We have mentioned only a few selected examples to characterize the work by Elion and Hitchings and their associates, and later, by Elion and her associates. We are by no means presenting an inventory of all the wonderful drugs they produced. The examples serve to illustrate that they did not merely produce these drugs—which in itself would have been a tremendous achievement—but they provided a whole new successful approach to drug design.

The year 1967 brought an important change for Hitchings and Elion. Hitchings was promoted to become the company's vice president in charge of research, and Elion became the head of the Department of Experimental Therapy. Increased responsibilities also meant increased possibilities for her, as well as more independence. Their joint work had been very fruitful, but there was always the possible misunderstanding among those who were not participants in their truly mutual cooperation that she was merely Hitchings's able assistant, rather than an equal contributor to the effort. She welcomed the opportunity to show what she was capable of doing on her own.

There was another change in their lives a few years later. In 1970, Burroughs Wellcome left Tuckahoe and moved to Research Triangle Park. The move created the possibility of considerable expansion of laboratory space and personnel, but it would not happen without some sad losses. Elion especially regretted that most of the dedicated young women in her laboratory could not make the move to North Carolina because of their family obligations. The difficulties proved temporary, and soon Elion and her growing team in North Carolina were back in business with higher intensity than ever. One of the success stories was their pioneering achievement in creating an antiviral drug.

Scientists used to think that antiviral drugs could not be manufactured because they could not be selective enough. Selectivity means being able to distinguish between the nucleic acids of the virus that are being attacked by the drug and the nucleic acids in the cells of the human organism the drug is supposed to protect. Elion and Hitchings had made some attempts at selectivity in the 1940s but had given up at the time. Elion always considered this failure temporary. By the late 1960s, she saw promising signs in the literature that encouraged her to revisit this abandoned area of their research. She now had better conditions, more staff, and modern instruments, so her team started producing purine derivatives that might be candidates for selective antiviral activity. They worked in close cooperation with the British team that had ongoing antiviral screening operations.

After long and painstaking work, Elion and her associates arrived at a compound called acyclovir, or ACV, which appeared suitable for treating the herpes-virus infection and moreover showed the necessary selectivity at the same time. Here, again, Elion's characteristic approach yielded benefits. Rather than just merely rushing ahead with developing a new drug, she devoted time and energy to understanding the reason for the high degree of selectivity of the new compound. This would provide useful knowledge about the biochemical differences between the interactions with the viral nucleic acids and the nucleic acids in the mammalian cells. Further, the team figured out the mechanism of ACV's action: As the virus attacks a healthy cell, it enters it, and once inside, it starts producing an enzyme that helps it multiply. If there is ACV present in the cell, as soon as the enzyme produced by the virus emerges, it triggers the transformation of ACV into another substance that destroys the virus. Thus the virus activity—in the presence of this drug—causes its own annihilation.

While these mechanistic studies were going on, Elion's team investigated the therapeutic potentials of ACV in animal model systems. Incidentally, however unpleasant herpes is, it may not be a life-threatening condition in an otherwise healthy organism. It can be fatal nevertheless when it attacks patients suffering from leukemia or cancer and patients who have undergone transplantation of organs or bone marrow. Under such circumstances, ACV could truly be a lifesaver. And, indeed, ACV became a great triumph for drug research and something more: it also became a huge commercial success for Burroughs Wellcome. In addition, the knowledge gained from working with ACV has proved of utmost value to further antiviral research. The lessons learned were soon utilized in working out the first drug against the AIDS virus, called AZT (azidothymidine). It came from Elion's former department, but after she had already retired.

Elion was not surprised when Hitchings's Nobel Prize was announced in October 1988, but she was genuinely taken aback when she learned that she was to share the honor. However, it would have been truly impossible to separate their contributions. The two shared the award with the British pharmacologist James W. Black; each of them receiving one-third of the prize. The citation did not mention specifically any of the drugs discovered by the laureates; rather, they were distinguished for their discoveries of important principles for drug treatment. This formulation gave added emphasis to the broad significance of their work.

Elion regretted only that her grandfather, her fiancé, and her parents could not see her success, but she could take comfort in knowing that by then literally millions had benefited from her efforts. Since she could not bring her own family to the Nobel ceremonies, she brought her nieces and nephews. She was happy sharing her joy with them, and she was made yet happier by the letters of gratitude and appreciation that flooded her mailbox after the Nobel announcement. The beneficiaries of her drugs learned for the first time about the person behind their lifesaving treatment.

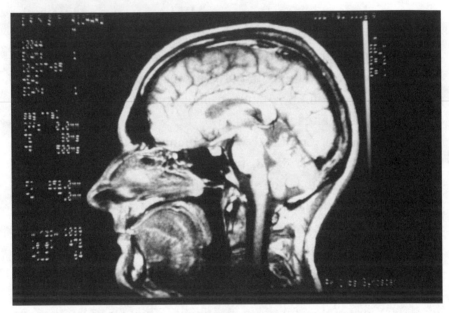

Figure 3.1. Magnetic resonance image of nuclear magnetic resonance pioneer and Nobel laureate Richard Ernst's head. Courtesy of Richard Ernst.

CHAPTER 3
OVERCOMING HANDICAPS
MRI

> [Our] disciplines [are] not natural
> categories . . . , but guides to instruction
> and efficient administration.
> Paul C. Lauterbur[1]

*Magnetic resonance imaging, or MRI, has become a standard
term in medicine, both for medical practitioners and for patients.
It is a noninvasive diagnostic tool that has helped hundreds of
millions. A Nobel Prize in Physiology or Medicine was awarded
for its discovery in 2003, and the awardees were a chemist, Paul
C. Lauterbur (1929–2007) and a physicist, Peter Mansfield
(1933–). Both took a delayed and unorthodox path in their education, although for different reasons. Besides them, many others
had also made crucial discoveries in related areas, but the Nobel
Prize was awarded for only the most crucial contributions to the
development of MRI. Hardly anyone expressed doubts that
Lauterbur and Mansfield deserved the distinction, but unprecedented controversies surfaced in regard to the merits of others.*

Peter Mansfield provided a terse description in his autobiography of leaving school, obviously prematurely, at the age of
fifteen. He had worked as a printer's assistant for three long years.
Then he obtained a job in the British Ministry of Supply at the

69

Rocket Propulsion Department in his native England, and he stayed there for a year and a half. At that point, he was called up to serve in the army for two years, so he returned to his previous job in 1954. In 1956—as his Nobel narrative continues—he entered Queen Mary College, University of London, and graduated in 1959.[2]

From this description I formed an image of a young man who had not realized the value of education at the proper time, and who took years to compensate for his error. I asked him about his youth when we met in 2005, and it turned out that he was no dropout; rather, the British school system of his time had failed him. He had to work very hard to become eligible for reentering the education system.

Mansfield went to an ordinary school without being aware of the existence of other kinds of schools. The story of his schooling is a little complicated but worth knowing about because Mansfield's road to science, let alone his milestone discovery, is an example of perseverance against odds. When he went to school, there were three levels of secondary school in England: the ordinary secondary school, the central school, and the grammar school. Pupils had to pass an examination to get into a grammar school. If pupils failed badly, they went to the ordinary secondary school; if they failed only mildly, they had the option of attending a central school. This was at the end of the war when Mansfield was eleven years old. He had just returned to London from evacuation and had to take the crucial examination within a week of returning. He was not prepared for it and did not get the requisite pass level, but he did not fail completely either, so he went to a central school.

At that particular time considerable changes were taking place in the structure of secondary schools. Many grammar schools were converted into so-called secondary modern schools, and the central school that Mansfield went to soon ceased to exist; hence, he ended up in an ordinary school. As a consequence, when at the age of fifteen his studies were completed in the ordinary school, he had no option to continue his studies! When the career adviser at his school asked him what he wanted to do, and Mansfield said he was interested in science, the interviewer burst into laughter. He found it ridiculous that a pupil of an ordinary school should aspire to become a scientist. He told Mansfield that he would never have a chance to become a scientist and suggested that he be sensible and tell him what he really wanted to be. As printing was one of Mansfield's hobbies, this is what he chose.

So he went into printing, initially as a bookbinder. Soon he switched his job and became a trainee compositor. Simultaneously, and from the very start, Mansfield attended evening classes. He was working in the City of London, and after work he rushed to the evening class at the Borough Polytechnic (today the University of the South Bank). He did this for five nights a week for three years until he passed the exams necessary to continue his studies at a higher level.

Mansfield was eighteen years old when he came across an article in the *Daily Mirror* about a boy of his age who managed to join a group of professional scientists working at the Rocket Propulsion Department in Westcott, Buckinghamshire. Mansfield wished he could do something like this and wrote to the newspaper editor, who suggested that he contact the Ministry of Supply. Mansfield did and received an invitation for an interview. The ministry, in turn, arranged an interview for him at the Rocket Propulsion Department, where one of the scientists offered Mansfield a position with the condition that he continue his studies. Though it may sound like a fairy tale, this is what happened.

When the time came, Mansfield took an examination, which was equivalent to the matriculation examination, and he passed it, so his job was secure. However, at this point he was called up to serve for two years in the army, and when he returned to civilian life in 1954, he continued studying, this time for university entrance. He studied three subjects to apply: physics, pure mathematics, and applied mathematics, and he also took chemistry.

He started his university education when he was twenty-three years old, in 1956, as a full-time student at Queen Mary College, University of London. His studies were supported by a grant from the local authority where he lived, the Buckinghamshire County Council. Then, within a few months, the Ministry of Supply invited him for an interview and, as a result, awarded him a scholarship. From that point on, everything, in Mansfield's words, "was all fairly straightforward because *I was driven.* . . . I was the only student who knew exactly what I wanted to do. . . . I was in a privileged position" (italics added).[3] Mansfield's eagerness to learn prompted him to take up a research project at the same time he did his undergraduate studies. One of the instructors, Jack G. Powles, was building a portable nuclear magnetic resonance (NMR) spectrometer to measure the magnetic field of the earth at various locations, and Mansfield became his student.

The utilization of nuclear magnetic resonance was relatively new, having been discovered and named in the United States two decades earlier. It came out of fruitful discoveries among several areas of physics. The primary discovery was an observation that the magnetic properties of atomic nuclei—themselves being tiny magnets—could be influenced by external magnetic fields. For the discovery of the resonance method of recording the magnetic properties of atomic nuclei, Isidor I. Rabi was awarded the Nobel Prize in Physics in 1944. A Nobel Prize was next awarded to Felix Bloch and Edward Purcell in 1952 for the development of techniques and for new discoveries in relation to magnetic resonance. Both Bloch and Purcell, with their associates, showed that magnetic resonance could be detected in bulk matter, not just in an atomic beam in a vacuum, as Rabi's experiments had shown. This was crucial for a cornucopia of applications that followed over the decades. The responses of the magnetic properties of atomic nuclei to the external fields revealed a tremendous amount of information about the internal structure of matter, at least to people capable of understanding their signals.

Upon the completion of his undergraduate studies, Mansfield became Powles's doctoral student and learned a great deal about NMR during the next three years. In 1962, at the age of twenty-nine, he received his doctorate, so he was finally on the track of scientific research. When I talked with him in 2005, he did not sound bitter about his late start. If he was unhappy about anything, it was that he might not have been decisive enough and had let others influence his fate too often. If he had been more decisive, he mused, he might have been twenty years younger at the time of our conversation. Realistically, though, he thought that a more assertive behavior might have upset things rather than advance them. He recognized that being stubborn was characteristic of him, and being yet more stubborn might have cost him much. Still, he felt, it might

Figure 3.2. Peter Mansfield in front of the Sir Peter Mansfield Magnetic Resonance Centre at the University of Nottingham, 2005. Photo by the author.

have helped him even more. He appeared to have a strong personality, as had Paul Lauterbur, whom I had met a year before.

Paul Lauterbur spent his childhood in Ohio. Practical considerations sent him to an engineering school, the Case Institute of Technology in Cleveland (which eventually merged with its next-door neighbor, Western Reserve University, to become Case Western Reserve University). Lauterbur graduated with a chemistry degree because, rather than fulfilling all the requirements for would-be engineers, he took additional courses in quantum chemistry. Although he was interested in research, he was not familiar with the trappings of a scientific career; he preferred actual work in the laboratory to sitting in more classes and was eager to do something useful. Thus, having earned his bachelor's degree, he joined the chemical laboratories at the Mellon Institute in Pittsburgh that belonged to Dow Corning Corporation, which subsidized some basic research at the Mellon Institute.

Lauterbur had his own special interest: silicon chemistry. He posed the general question of whether a silicon world—similar to the carbon world—could be developed. He was also keen on learning new techniques, and although the Mellon Institute had no NMR machine at that time, he gradually picked up a working knowledge of the emerging field of NMR spectroscopy. One of his sources was a lecture by a visiting scientist, Herbert S. Gutowsky, of the University of Illinois at Urbana–Champaign. Gutowsky was a leading scientist in liquid NMR techniques and pioneered their applications to chemistry and biochemistry. It seemed as if the two might have developed a rapport, but the relationship came to naught when Lauterbur was drafted for military service.

Lauterbur was directed to the medical laboratories of the Army Chemical Center in Edgewood, Maryland, which were involved in studies of chemical warfare. Researchers there hoped to produce fluorine compounds that would be nastier than nerve gases, but what they produced turned out to be inert. They worked with a plethora of other substances that had the potential of becoming a poison on the battlefield. They worked on body armor, hung it on goats, and shot at the goats as a test. As if fate would direct Lauterbur toward NMR spectroscopy, he had the opportunity to work with an NMR machine, which another group had acquired but which had no one to operate it. A fast learner, Lauterbur became a self-made specialist of NMR spectroscopy. By the time his service ended, he had produced four publications on nonclassified studies.

When Lauterbur returned to the Mellon Institute, he used the recently acquired NMR machine extensively for his research. He soon understood that since he lacked a PhD degree, some opportunities would stay closed to him. So, based on the work he had done, he developed his PhD dissertation and defended it at the University of Pittsburgh. By then he was thirty-three years old.

Lauterbur had a broad and innovative overview of the NMR field. His calculations showed that silicon could also be used in magnetic resonance experiments just as carbon could, and that the silicon-29 nuclei would be the silicon isotope to be utilized. His closer look at carbon NMR spectroscopy led to one of the first publications on carbon-13 NMR spectroscopy, which was followed by an explosive development of the field.

There were always people skeptical of Lauterbur's innovations, which is typical when a researcher runs too fast ahead of the pack. And perhaps his low-key style was not always sufficiently impressive. In some years time, by the time his interest in medical applications had developed, Lauterbur described to one of the authorities of the NMR field what he was trying to do for NMR imaging. This person told him that it either would not work or, if it did work, Lauterbur would not be the one accomplishing it, because he did not have it in him. However, Lauterbur would not be shaken from his chosen path at this time or ever.

Lauterbur took a position as associate professor at the State University of New York (SUNY) at Stony Brook after earning his PhD degree in 1962. Thus Lauterbur had never been a full-time graduate student; had never had a research adviser; had never held a postdoctoral position; and had never been an assistant professor. His academic career was straightforward from this point, and soon he was a tenured full professor. He set up an NMR laboratory at Stony Brook and continued his wide-ranging research parallel to his pedagogical activities. He spent his first sabbatical leave at Stanford Medical Center of Stanford University and made numerous contacts in California, including Syntex Company and Varian Associates, both heavyweight firms in the NMR field. It was at this time that Lauterbur started gradually developing an interest in medical applications of NMR.

The story of the beginnings of NMR's medical applications is a complicated one and involved many scientists in numerous laboratories. When a paper about the beginnings of nuclear magnetic resonance imaging appeared in a popular-science magazine in 1992, it was praised for objectivity but immediately augmented with several additions by similarly objective contributions.[4] Our narrative does not attempt anything comprehensive in this respect; it merely mentions a few road posts on Lauterbur's way to his discovery and its afterlife. There were many sources of Lauterbur's interest in the medical applications, and one of them served as an immediate spark: it was a fledgling company with which Lauterbur had become involved.

A former Varian engineer had started a company near Pittsburgh called NMR Specialties, Inc. Lauterbur had first served as a consultant for the company, then became a member of the board of directors. When a crisis had developed one early summer, it seemed that the company would cease to exist unless it found someone to take over. Lauterbur was free from his academic duties for the summer, so he was elected to this post.

The company had an arrangement whereby specialists could come and use its NMR machines for their research. This is how Raymond Damadian came to test his revolutionary idea about distinguishing between healthy and cancerous tissues of rats by the differences in their NMR spectra. He measured relaxation times; that is, the lengths of time the magnetic properties returned to pre-experiment conditions after he had stopped applying external magnetic field to his samples. Damadian realized that NMR might be useful for the detection of tumors, and that the relaxation times appeared different for the healthy tissues and for the malignant tissues. He published his observations in the spring of 1971.[5] He did not use a large number of samples, just six normal specimens and two with malignant tumors—each with two different kinds of tumor. Damadian reasoned that NMR imaging might be advantageous for early detection of tumors as compared with X-rays, for which many tumors showed high permeability (thus remaining undetected), especially at an early stage of their development.

Soon after, an NMR spectroscopist at Johns Hopkins University came to use the same facilities, bringing with him a bunch of rats. He was looking for confirmation of Damadian's findings. Clearly, something was brewing in directing the potentials of NMR spectroscopy toward medicinal applications. Unfortunately, all these experiments necessitated either killing animals or taking heavy invasive actions, and in my conversation with Lauterbur in 2004, he expressed his distaste for this aspect of the early trials. He spoke about his "concern for sacrificing animals in operating on them."[6] It would be a mistake though to ascribe Lauterbur's interest leading to MRI primarily to his concern for animal welfare. It was merely "a psychological trigger," but not a "real motivating force."[7]

Lauterbur realized that something very different from previous approaches to imaging was possible. He thought hard, and an idea occurred to him that was different from any other he had had before—one with the promise of wide applications. The idea came to him while he was having dinner with one of his NMR colleagues, Don Vickers. (This was such a memorable event for Lauterbur that when his Nobel Prize was announced, he invited Vickers to join him at the Stockholm celebrations.) Lauterbur's basic idea was to employ external magnetic fields of varying strength for producing images of the target under investigation and reconstruct its two-dimensional image. Thus, images of slices of the target could be re-created, and then from the slices a three-dimensional picture could be reconstructed.

Experience in the industrial laboratory of Dow Corning had taught Lauterbur that he should record his thoughts and validate his notes. As soon as he realized that he was having new ideas about imaging, he bought a notebook in a nearby drugstore, and during the night he wrote down his idea in general terms. The next morning, he had his notes witnessed for potential patent issues. He was seriously considering filing for a patent and was working on it for a while with a patent attorney at the company. However, before they could complete the application, they had a falling-out over an unrelated business matter.

At that point, Lauterbur went to his university, SUNY, because he judged that even a percentage of whatever came out of such a patent would be better than nothing. The university employed an organization for working on its patents, but it was decided that Lauterbur's idea would not be a viable patent worth its effort and expense. The university was wrong by many orders of magnitude, but at the time, Lauterbur got discouraged and abandoned the idea of patenting. He regretted in retrospect that he did not pursue the idea of patenting, not so much because of the monies he did not earn but because the existence of patents—however paradoxical this may sound—would have accelerated industrial applications.

As soon as Lauterbur's thoughts crystallized, he prepared an

Figure 3.3. Paul C. Lauterbur in 2004 in the Lauterburs' home in Urbana, Illinois. Photo by the author.

article for publication and submitted it to *Nature*. He was so fascinated with his new general principle of imaging that he left out the possible utilization of his approach for cancer detection and medical diagnosis; instead, he concentrated on basic science. *Nature* rejected his manuscript right away and without any specific justification. It just did not see the novelty in Lauterbur's idea and gave it a blanket rejection. At this point, Lauterbur felt he had to be stubborn. He wrote *Nature* a long, impassioned appeal and offered to expand his manuscript to make it more convincing, describing its potential for applications. Later he learned that his new manuscript was sent out to a different reviewer who thought that the manuscript seemed crazy but that Lauterbur was known not to have done anything crazy before. To the credit of the magazine, it then published his paper.[8] Many years later, *Nature* included it among the twenty most influential papers published during the previous one hundred years. This was quite a distinction, because *Nature* publishes many pioneering papers.

Something similar to the fate of his *Nature* publication happened to Lauterbur's proposal when he applied to the National Institutes of Health for support for his research project. Again, as Lauterbur learned about it later, the first reviews were negative; the reviewers found his proposal insane. Then somebody on the committee suggested taking a second look at the proposal exactly because it was so unusual, so they did. The result was that the committee members still found it crazy, but they could not find anything specifically wrong in the proposal, so they decided to fund it. When there is a truly original proposal, there are not yet any real peers to judge it. According to Lauterbur, "The experts are very proud of their expertise and disappointed when something proves to be original that they had not thought of themselves and they like to look for flaws and tend to think that it may not be true; there may be all sorts of psychological reasons."[9]

When patenting was no longer an option for Lauterbur, he made much use of the freedom and openness that were possible for him in the absence of patenting. As he told it, he made virtue out of necessity and started promoting his ideas of imaging. He went nearly everywhere giving lectures and invited everybody to see what he had in his laboratory. One of the consequences of Lauterbur's interest in the biological applications might have been his meeting Joan Dawson in 1978 at Oxford University. She had been involved in doing NMR physiology in vivo, although not with imaging. Later, she became Lauterbur's second wife and professor of molecular and integrative physiology, biophysics, and neuroscience at the University of Illinois at Urbana. As for patenting, Lauterbur thought that the awarders of the Nobel Prize probably appreciated that he did not patent his discoveries. This is not necessarily so, because Peter Mansfield was very successful in patenting his ideas, and this did not seem to bother the Swedish judges.

After many years of working with NMR, Mansfield started thinking about imaging in 1972.[10] He had been interested in applying NMR spectroscopy to solid samples, but in the academic year 1975–76, while he was spending a year in Germany at the Max Planck Institute for Medical Research in Heidelberg, the medical environment may have induced him to switch to liquid samples. The liquid state is closer to living organisms than solids or gases: living organisms are full of water. Mansfield went into the garden and took some flower stems, twigs from trees, anything alive that was small enough to put into the NMR machine. Then a student put one of his fingers into the machine. This was the first time they had the image of a live human body part.

By then, Mansfield had been associated for some time with the Physics Department of the University of Nottingham in England. There was some rivalry in working out the imaging possibilities by NMR because the head of the department, Raymond Andrew, and his research group were also doing research along the same lines. Both groups were being funded. The competition may have been stimulating, but it also made the participants' lives unpleasant. Mansfield's situation improved when he was appointed professor in 1979. The period between 1976 and 1984 was his most productive. In the late 1970s, his department was ready to build an NMR machine large enough to hold a whole human body. Alas, they were not the first. Raymond Damadian had already accomplished that a few months before.

In 1984, all of Mansfield's competitors at Nottingham left for the United States. He almost went to America too, because there was not sufficient support for his work in the United Kingdom. He and his family were ready to move when, at the last moment—and he does not know whether this was coincidence or by design—he got word from the British Department of Health that it had suddenly found the monies to support the continuation of his ambitious project. Ten years later, at the age of sixty-one, Mansfield retired from the University of Nottingham to devote himself full-time to the perfecting of MRI.

Lately, Mansfield has been involved in trying to solve some outstanding problems in the medical applications of NMR. This has been so much on his mind that he mentioned it in his Nobel Banquet speech in Stockholm (the banquet speech is usually about highbrow ideas and generalities rather than about unsolved research problems). Mansfield has been keen on finding a way to eliminate the clanging noise of the MRI machine, which can be very disturbing for patients undergoing examinations. He set up a small company with just two employees, and, thanks to considerable royalty income during the preceding decade from his patents, Mansfield had the necessary resources to do so. Had he and his team solved the problem, the solution might have led to new designs of MRI machines, which could be very lucrative; alas, the solution has so far eluded them. Mansfield complains that due to his character he is unable to give up on a problem once it captures his mind. He keeps thinking that maybe tomorrow he will solve it.

In contrast, Lauterbur considered his excursion into MRI a detour from his main interest in chemistry, even though the detour lasted thirty years. He changed his main direction of research back to chemistry from medical imagery well before the Nobel Prize. He became interested in the chemical origin of life on earth. When he received the Nobel Prize, he was still determined to go on with his interest and not allow the hoopla around the big award to disturb his routine too much. Sadly, he did not live long after receiving the award. I visited him in Urbana, Illinois, in February 2004; we met again in June 2005 at the meeting of Nobel laureates in Lindau, Germany, by which time he was in a wheelchair. He died in the spring of 2007. Considering that Lauterbur lived a mere four years after the award, the Nobel Prize for magnetic resonance imaging came almost too late. Richard Ernst, the chemistry Nobel laureate in 1991 for his NMR-related discoveries, lamented about the omission shortly before the 2003 award was announced. He declared that "[t]he disrespect for MRI in Stockholm is particularly difficult to understand."[11] These were strong words from a Nobel laureate; Nobel laureates seldom criticize Stockholm.

The story of the 2003 Nobel Prize was colored by an unusual protest. After the Nobel Prize was awarded to Paul Lauterbur and Peter Mansfield, Raymond Damadian took out whole-page advertisements in the *New York Times*, the *Washington Post*, and the *Los Angeles Times* in the United States and in a leading Swedish newspaper, protesting the decision that had denied him the prize. Protests about Nobel Prizes are not rare, especially when worthy scientists are left out; what was unusual in this case was that the overlooked scientist himself waged the protest, and it was a very forceful protest at that. As any Nobel Prize can be shared by up to three people, there was an empty slot that could have been used for Damadian, but it could have been used for others, too. There are opinions that, besides Lauterbur and Mansfield, it was not at all straightforward who the next in line might have been. Obviously, the Swedish judges preferred to leave the spot unused rather than make a hard choice. There has never been a change in Nobel Prize decisions once they have been made, and there is no authority higher than the prize-awarding institutions that could make the prize awarders change their published choices.

Damadian's pioneering role cannot be denied, and his NMR machine—called Indomitable—is a piece of medical history. As such, it is stored at the Smithsonian Institution in Washington, DC. Among Damadian's distinctions are the National Medal of Technology (1988) by the president of the United States and the Bower Award for Business Leadership (2004) by the Benjamin Franklin Institute in Philadelphia. His contribution has also been recognized by the US Supreme Court in the outcome of a patent dispute against General Electric in which his company, FONAR, was awarded over one hundred million dollars.

In my conversations with the two laureates, Lauterbur's principal response to my queries was a copy of Donald Hollis's book titled *Abusing Cancer Science*, whereas Mansfield told me in some detail about his rather disappointing experience with Damadian.[12] Mansfield stressed: "Everyone seems to feel that number one should be Paul Lauterbur."[13]

There is one more aspect of the MRI story that even a brief account should touch on, and that is nomenclature. Coining names is an integral part of a discovery; according to some, it is the only genuinely original contribution by the discoverer for his or her recognition. A discovery—in contrast with creations in the arts—is made sooner or later by one scientist or another, but coining a name for it makes it truly personal. In this particular case, however, this name-giving preference did not work. Damadian called his machine Indomitable and his company FONAR. Lauterbur used the terms *zeugmatographic image*, *zeugmatographic principle*, and *zeugmatographic techniques* based on *zeugmatography*, derived from the Greek *zeugma*, meaning "that which is used for joining," referring to the fact that MRI came together from joining different areas of science.[14] But none of these terms have caught on, whereas *MRI* has.

There is, of course, a difference between NMR imaging and MRI: the word *nuclear* is conspicuously missing from the latter.

The change came sometime in the mid-1980s and may be in part a result of squabbling between various American organizations representing NMR people and nuclear medicine. However, no one denies that the term *nuclear* was gladly dropped by those who were concerned about the popular image of this diagnostic tool and the effect that anything labeled "nuclear" has on the public. The name change would not alter anything regarding the nuclei that produce the effects utilized in this tool. But we have all learned that perception counts. Incidentally, MRI does not produce any harmful radiation, but it may interfere with pacemakers.

In the preceding narrative, I did not mention many of the other important contributors to the development of MRI, and I restricted my discussion to Lauterbur, Mansfield, and, to a small extent, Damadian. Perhaps a further caveat is warranted. I spoke about Lauterbur's and Mansfield's careers, stressing how they overcame their handicaps on their way to each becoming one of the world's most distinguished scientists. Their careers could also be looked at from a different viewpoint. We could see in them not only their own resilience and perseverance but also the flexibility of the systems within whose frameworks they operated and under which they could fulfill their dreams. It is mind-boggling how many great talents must be lost all over the world that are under much more severe conditions than were accorded to our heroes. It is to their merit and to the merit of the environments in which they operated that they accomplished what they did—to the benefit of us all.

Figure 4.1. Ava and Linus Pauling. Photo by and courtesy of Karl Maramorosch, Scarsdale, New York.

CHAPTER 4

SPIRIT OF COMPETITION
Structure of Proteins

. . . great progress will be made, through this technique [structural chemistry] in . . . biology and medicine.
Linus Pauling[1]

Linus Pauling (1901–1994), one of the greatest chemists of the twentieth century, never publicly admitted that there was a race for the determination of the structure of the most important biopolymers. But according to his competitors there were two races, and Pauling won one (the structure of proteins) and lost the other (nucleic acids). He had a tremendous number of ideas, many of which he deemed worthless, but a few were brilliant. Not only did he make seminal discoveries, he was also a master in announcing them. Eventually, Pauling shifted toward politics and controversial issues, but his science ensured him his place among the greats. Here we follow Pauling's route to the discovery of the alpha helix and the defeat of the star-studded British team in the same quest. We also mention the fate of the theory of resonance that ensured his victory in his quest for the protein structure. Yet for this same theory of resonance, the politically leftist Pauling was strongly condemned in the Soviet Union.

The British father-and-son team of W. H. Bragg and W. L. Bragg pioneered the technique of X-ray crystallography in 1913, and when they were awarded the Nobel Prize in 1915, the son, at age twenty-five, became to that date the youngest ever Nobel laureate. After a hiatus due to World War I, this field took off spectacularly in the realm of small molecular systems. A silk fibroin, which was supposed to be a macromolecular system (a polymer), was exposed to X-ray diffraction for the first time in 1928 by Herman F. Mark in Germany. Polymers are virtually limitless chains of repeating small-molecule units. In this experiment, a beam of X-rays was shone onto a sample of silk fibroin, and the way it scattered the X-rays was supposed to give information about its structure. Mark was to become one of the century's foremost polymer chemists. When he was forced out of Germany, he moved to his native Vienna and helped one of his students, Max Perutz, to become a doctoral candidate in Cambridge, United Kingdom, in 1935. There, Perutz would become a key player in the quest for the structure of proteins. At the time that Perutz was still learning the basics of this science, Linus Pauling was already a major force in the field.

Pauling came from a modest background, but he had drive. He lost his pharmacist father when he was nine years old, and his mother found it difficult to cope with her obligations. Little Linus read the Bible, Charles Darwin's *Origin of Species*, and most of the *Encyclopedia Britannica* by the age of twelve. In 1922, Pauling went to the California Institute of Technology (Caltech), which, while not yet the great institution it would become, was as ambitious as its new student. The visionary movers of Caltech who were to develop it into a top-notch institution of higher education and research recognized in Pauling the potential of a star scientist.

For his doctoral studies, Pauling became engaged in the determination of the structure of a variety of inorganic and organic molecules and amassed a large amount of information about them during the ensuing decade. However, not all the modern knowledge was to be had at Caltech at the time, and Pauling—like many other aspiring American scientists—paid pilgrimage to a series of European research centers to learn from the likes of Arnold Sommerfeld in Munich and Erwin Schrödinger in Zurich. They were both physicists, but Pauling's aim was not to transform himself into a physicist. Rather, his aim was to apply the latest discoveries in physics, especially the new quantum mechanics, to solving problems in chemistry, and in this, he proved to be unique.

The most intriguing question in chemistry was about the forces that keep the atoms together in a molecule; that is, about the nature of the chemical bond. This area of research would become Pauling's trademark. An intuitive approach is often required in chemistry, stemming from a desire to represent on paper what is experienced in the laboratory. Thus, for example, chemists started using a straight line connecting the symbols of two elements to represent their bonding without yet understanding what that straight line truly meant. Nowadays, despite all we've learned, we still find this straight line to be an excellent representation of the chemical bond.

The new physics, with its more rigorous approach, helped build a foundation for understanding chemical bonding, but it was too sophisticated and too narrowly applicable for chemists to benefit from it. Pauling bridged this gap in a series of brilliant articles in the *Journal of the American Chemical Society*. Eventually, he developed his ideas and his repository of structural information into a bestselling book, *The Nature of the Chemical Bond*.[2] Its third and last edition appeared in 1960, and many of the later stars of chemistry benefited by getting their introduction to the intricacies of their science from this book. Though an updated book on the subject would be timely, nobody seems brave enough to try filling Pauling's shoes by producing a new, comprehensive monograph about the chemical bond.

Had Pauling produced only his series of articles and his book about the chemical bond, he would have already written his name into the annals of chemistry. However, he did not restrict his interest to theoretical studies. He utilized X-ray crystallography broadly and was constantly on the lookout for new techniques. In Germany, Pauling visited Herman Mark's industrial laboratory, where Mark showed him a new technique—gas-phase electron diffraction—for the determination of molecular structure using an electron beam, which fortunately augmented X-ray crystallography.

Mark did not plan to utilize this new technique, since it was not applicable for practical purposes, so he encouraged Pauling to take it with him back to Caltech. Mark even supplied him with the blueprints of his apparatus. Pauling not only quickly introduced the new technique in the United States, but he and his student, Lawrence Brockway, further developed it. This was another example of Pauling's very broad talents, which ranged from pure theory to novel experimental work. During most of my own research career, I have used the electron diffraction technique, and one of my mentors, the Norwegian Otto Bastiansen, did his postdoctoral studies with Pauling; thus, to some extent, I am a descendent of the Pauling School.

Pauling established relationships among various experimental facts and made predictions about structures not yet investigated. He then worked out a theoretical technique based on quantum mechanics —simple enough for a broad circle of chemists—to describe molecular structures. It was called the valence-bond (VB) theory, and it was one of the two major theoretical approaches developed over the decades. For chemists, Pauling's theory had a strong appeal.

An important feature of the VB theory was that a molecular structure could be described by a set of interconverting; that is, "resonating" structures, and the sum of these resonating structures corresponded to the real structure. One of the simplest examples of the resonance description can be given for the benzene molecule

C_6H_6. This molecule consists of a ring of six carbon atoms with a hydrogen atom linked to each carbon atom. Simple chemical considerations excluded that the six carbon–carbon bonds along the ring would all be exclusively either single bonds or double bonds. The solution appeared to have three single bonds and three double bonds alternating in the ring. However, there are two possible ways to draw such structures.

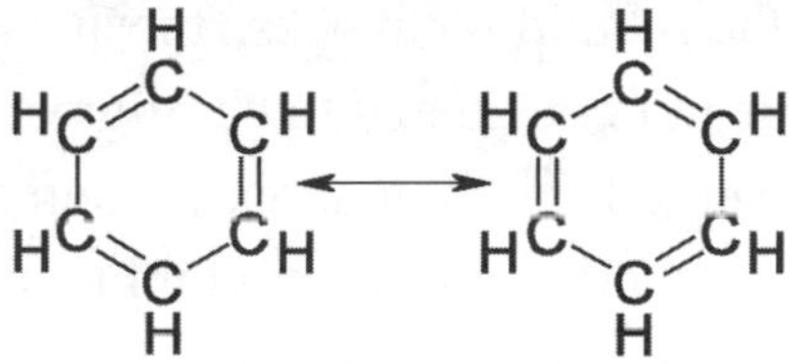

Figure 4.2. "Resonating" models of the benzene molecule.

These two drawings are each other's mirror image, but otherwise they are equivalent. So which is the right one? The resonance approach solves this puzzle by suggesting a quick interconversion —resonance—between the two structures. The result is an average of the two extreme models, and in the average there are no alternating single and double bonds, but all bonds are equivalent and intermediate between the single and double bonds. Such a description finds excellent experimental support, because when the distances are measured between the carbon atoms, they are all of equal length, somewhere in between those that would indicate either a single bond or double bonds.

There were proponents and opponents of the theory, as is the case with most theories. Pauling was one of the initiators of the resonance theory in chemistry, and it proved decisively useful for him in his quest for the protein structure. This theory showed him the way to proceed and brought him a resounding victory over his competitors who lacked this tool and could not arrive at the right solution.

Pauling advanced in a systematic manner in his quest for building up structural chemistry. First, he busied himself with inorganic substances, and after spending ten years on that endeavor, he moved to organic substances. As Pauling was learning more and more about the structures of relatively simple molecules, it occurred to him in the mid-1930s that he might as well make an attempt to learn about larger molecules. Proteins were an obvious choice, because they were the most important biopolymers. Although nucleic acids were also known, and their building blocks, the nucleotides, had been identified, they were not considered to be of great significance.

When Pauling started thinking about protein structures, the first protein to attract his attention was hemoglobin—the vehicle that carries oxygen in an organism. Incidentally, Max Perutz of the British Cambridge group, who was also engaged in protein structure studies, had selected hemoglobin for his target as well; his choice was independent of Pauling. W. L. Bragg was the research director in crystallography in Cambridge, and Perutz would be joined by John Kendrew after the war. Pauling formulated a theory about the oxygen uptake of hemoglobin and the structural features of this molecule related to its function of disposing of and taking up oxygen.

Pauling's interest in protein structures was further excited when a visiting scientist and protein specialist, Alfred Mirsky of the Rockefeller Institute, spent the academic year 1935–36 in his laboratory. They jointly studied the behavior of proteins under the impact of heat or chemical substances. They described proteins as having a regularly folded structure in which hydrogen bonds provided the stability of the structure. Hydrogen bonding was a recently discovered phenomenon. It was gradually becoming recognized as a crucial mode of interaction in chemical structures and especially in those of biological importance. In retrospect, it was a pivotal discovery, but one whose significance was not at all obvious when it was first described; it only emerged slowly over the years.

Hydrogen bonding is weaker than the normal chemical bond, and one of its great features is that it can be established and disrupted more easily than normal chemical bonds. For many biological molecules, it is the hydrogen bonds that keep their different parts together. Hydrogen bonding is attributed to the partial positive charge on the opposite side of the hydrogen atom from the side where its bonding to another atom occurs. The other partner in hydrogen bonding is an atom, for example, oxygen, which in addition to its participation in normal bonding possesses electrons—as if representing a negative charge—not involved in the normal bonding. The attraction between

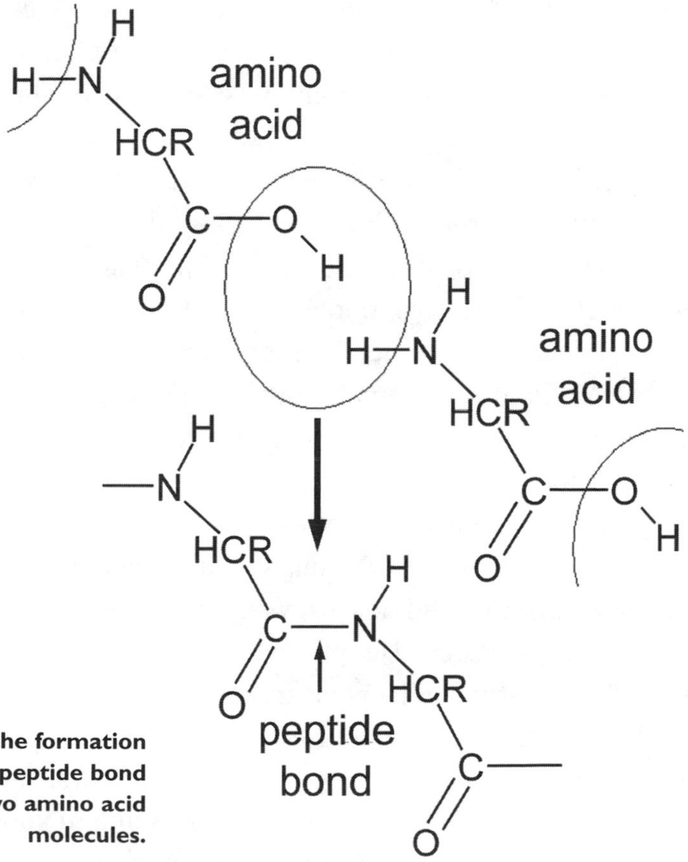

Figure 4.3. The formation of a peptide bond between two amino acid molecules.

the positive side of hydrogen and the negative side of oxygen (or another atom) establishes the hydrogen bonding.

The building blocks of proteins are the amino acid units, which are linked to each other by a peptide bond. The peptide bond is formed by a carbon atom of one amino acid and a nitrogen atom of the next amino acid, and so on. Thus a protein molecule is a chain of amino acid units linked by peptide bonds.

Pauling further postulated that the subsequent amino acid units are linked to each other in the folded protein molecule not only by the peptide bond but also by a hydrogen bond that occurs because the folding brings the participating atoms sufficiently close to each other. When the protein molecule then unfolds, these hydrogen bonds break. Because practically nothing was known about the nature of folding, this was only a hypothesis. Finding out more about it occupied Pauling's mind for the next fifteen years.

By the time Pauling became engaged in this research, it had been established from rudimentary X-ray diffraction patterns that there might be two principal types of protein structure. Keratin fibers, such as hair, horn, porcupine quill, and fingernail belonged to one, and silk to the other. The foremost British crystallographer of fibers, William T. Astbury, showed in the early 1930s that the diffraction pattern of hair underwent changes when it was stretched. He called the protein structure that produced the normal pattern alpha keratin and the other, which was similar to the pattern from silk, beta keratin. In 1937, Pauling set out to determine the structure of alpha keratin. He did not just want to rely on a single source of information, however. He planned to use all his accumulated knowledge in structural chemistry and find the best model that would make sense and would be compatible with the X-ray diffraction pattern. This comprehensive approach was advantageous to him and was obviously drawn from his wealth of knowledge. He was a walking data bank of structural chemistry.

One helpful piece of information from X-ray diffraction was that a certain part of the protein chain was being repeated along the chain at every 5.1 angstroms (= 0.51 nanometers = 510 picometers). Another piece of information was the length of the peptide bond, which indicated that this bond was not purely a single bond, but it was not purely a double bond either. Pauling's involvement with the resonance theory told him that the emerging structure could be represented by two resonating structures.

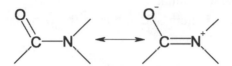

Figure 4.4. "Resonating" peptide bond.

From the accumulated structural information he also knew that the bonds around a double bond are all in the same plane. This was a crucial piece of information because, rather than taking into account all kinds of rotational forms with respect to the peptide bond, he could assume that all the atoms around the peptide bond were in the same plane; in other words, the peptide group was planar. This assumption greatly reduced the number of possible models he had to consider for describing the structure of alpha keratin: rather than a virtually infinite number of possibilities—forms of rotation about the peptide linkage—he could choose just *one*. Alas, even with all this knowledge, Pauling at this time was unable to find a model that would fit the X-ray diffraction pattern, and he postponed further work on protein structures.

During the ensuing years, Pauling and his newly arrived associate, Robert Corey, an expert in X-ray crystallography, determined the structure of a large number of amino acids and simple peptides. At some time every doctoral student in Pauling's laboratory was supposed to determine the structure of an amino acid for his PhD

dissertation. Pauling managed to persuade his students to participate in this program without directly ordering them to do so. When
the noted Harvard biochemist Matthew Meselson was Pauling's
graduate student, he was also given a choice between an amino acid
or a very different project, like a tellurium compound. Their conversation went like this:

Matt, do you know about tellurium?

Yes, Professor Pauling.

It's under selenium and sulfur in the periodic chart of the
elements.

Yes, Professor Pauling.

Have you ever smelled hydrogen sulfide?

Yes, Professor Pauling.

It smells bad, doesn't it?

Yes, Professor Pauling.

Have you ever smelled hydrogen selenide?

No, Professor Pauling.

Well, it's much worse.

I see, Professor Pauling.

Now, Matt, have you ever smelled hydrogen telluride, you
probably have not.

No, Professor Pauling.

It's much worse than hydrogen selenide.

I see, Professor Pauling.

If you want to work with the crystal structures of some salts
of tellurium, I want you to be very careful because some chemists
had gotten tellurium into their system, and they acquired something called tellurium breath. It would isolate you from society
because it's so bad. Some of these people have committed suicide. So what do you think?[3]

Meselson opted for an amino acid, as did many others. He never found out whether or not Pauling was serious in offering a choice of research topics, but he found him to be a wonderful teacher.

Meselson was one of a large group of Pauling's graduate students after World War II who all went on to brilliant careers. The war had caused a delay in Pauling's structural work, as he was engaged in war-related research. After the war, however, Pauling and his team vigorously resumed their quest for the protein structure. Corey and Pauling complemented each other famously as researchers. Pauling had excellent instincts and intuition and was never reluctant to come to firm conclusions with a minimum amount of data. Corey was the exact opposite, always wanting to be absolutely sure. Pauling could create a coherent manuscript during a single dictation; Corey would fuss with every word.[4]

Pauling became intensively engaged in protein research in 1948 while he was a visiting professor at Oxford University in England. By then, the amount of experimental information he had about the building blocks of proteins greatly expanded. Furthermore, it had the beneficial effect that Pauling could now take a more detached view of the problem in his renewed efforts. When he was looking for the solution more than a decade before, he was bothered by the knowledge that his model was supposed to accommodate the possible presence of twenty different amino acids in the protein chain. At this time he decided to ignore their differences and assumed them to be equivalent for the purpose of his model. This was yet another example of Pauling's ability to distinguish between essential features and those that could be ignored in building his models.

Pauling's broad-based education also paid off. He remembered a theorem in mathematics he had learned a quarter of a century before, which stated that the most general operation to convert an asymmetric object into an equivalent asymmetric object is a rotation–translation, and that repeated application of this operation produces a helix. *Rotation–translation* means a simultaneous turn around and shift along the axis of the molecule. Here the asymmetric objects were the amino acids constituting the protein chain, and the amount of rotation was such that it took the chain from one amino acid to the next. And during all these operations the peptide group was kept planar. An additional restriction was keeping the adjacent peptide groups apart at a distance that corresponded to hydrogen bonding.

In Pauling's model, the turn of the protein chain did not involve an integral number of amino acids—he did not consider this a requirement, whereas his British counterparts did. This was yet another relaxed feature of the structure that served him well, whereas insisting on having an integral number of amino acids at every rotation proved to be an unnecessary restriction for his competitors.

For Pauling, the moment of discovery came when he was confined to bed with a bad cold. Pauling—ever the model builder—sketched a protein chain on a sheet of paper, which he folded while looking for structures that would satisfy the assumptions he had made. He found two. He called one the alpha helix and the other the gamma helix, but he would quickly discard the gamma helix.

We have already seen that the repetition along the peptide chain occurred every 5.1 angstroms, according to the X-ray experiments. To Pauling's disappointment, when he measured his rudimentary model, it indicated a 5.4-angstroms repeat distance (rather than 5.1). However, Pauling found the model so attractive and so sensible that he had little doubt in its correctness. Nonetheless, he decided to postpone publishing his findings until he could understand the discrepancy. At this time, he had an opportunity to visit the British group in Cambridge, and Max Perutz showed him his excellent diffraction patterns. From the X-ray diagrams it was obvious to Pauling—though not yet to Perutz—that it corresponded to Pauling's alpha helix model. Pauling was excited by what he saw but kept calm and did not say anything to Perutz.

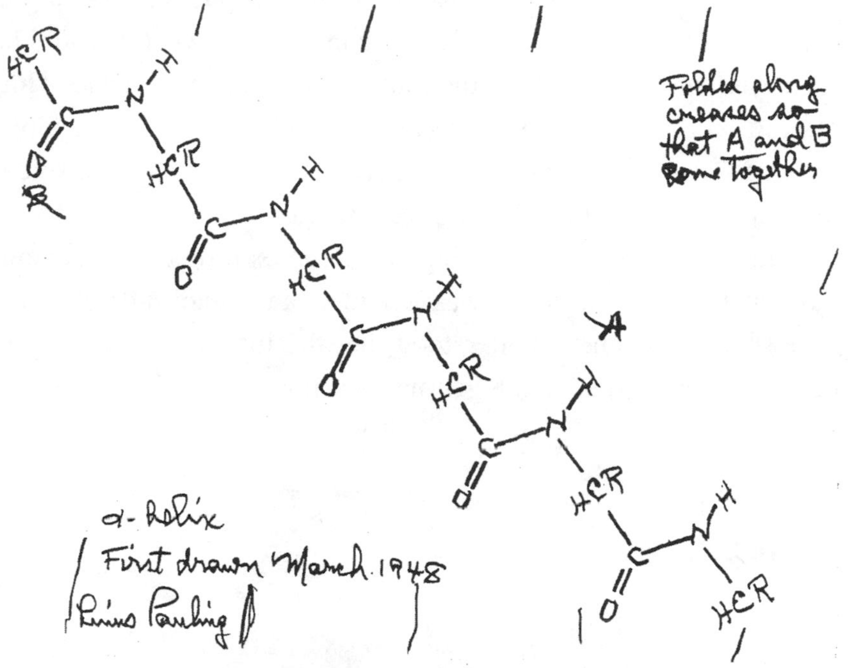

Figure 4.5. Linus Pauling's sketch of the polypeptide chain in 1948. When he folded the paper along the creases, the alpha helix appeared. Linus Pauling, "The Discovery of the Alpha Helix," *Chemical Intelligencer* 2, no. 1 (1996): 32–38, with kind permission of Springer Science+Business Media.

When Pauling returned to Pasadena, he and his associates double-checked all his calculations and found no errors. About a year passed, and the British team—Bragg, Perutz, and John Kendrew—published a long article about protein structures and communicated about twenty models, none of which contained a planar peptide group, and none of which described alpha keratin satisfactorily.[5] Finally, Pauling made up his mind to ignore the discrepancy of the repeat distance between his model (5.4A) and the experimental observation (5.1A). He and his associates published their alpha-helix model.

Eventually, the origin of the discrepancy was understood: it was caused by the alpha helices twisting together into ropes resulting in a change in the experimental data as compared to what it would be for a single chain for which the model had been constructed. Thus Pauling's alpha helix was confirmed even in this detail. The alpha helix has proved to be a great discovery because it is a conspicuously frequent structural feature of proteins.

Pauling's approach to solving this complex problem was exemplary in focusing on what was essential and ignoring what was nonessential. When he understood that the turn around the chain did not correspond to an integer number of amino acids—hinting at

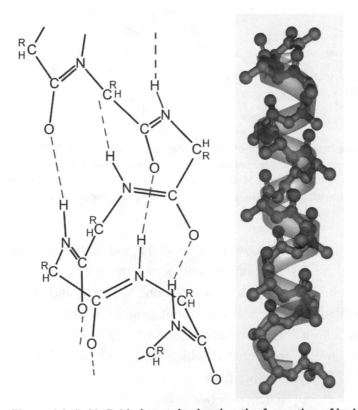

Figure 4.6. (left): Folded protein showing the formation of hydrogen bonding within the helix. This drawing indicates the double-bonding character of the peptide bonds. Figure 4.7 (right): Computer drawing of the alpha helix. Courtesy of Ilya Yanov, Jackson, Mississippi.

less than perfect symmetry—he did not let himself be bothered. He simply expanded the realm of crystallography toward structures that were not part of classical crystallography yet were included in literally vital substances.

It was also interesting that he could skip a decade working on this most important discovery without much danger of being scooped. Others almost did scoop him, but their only advantage was their timing, and not their knowledge, because Pauling's knowledge proved to be superior to that of anyone in the field at that time. However, when he visited the Cambridge group, he could sense that he was no longer alone, and from that point on, the time factor must have entered his consideration. The Cambridge X-ray diffraction pattern showed the helical nature, but Perutz did not think about it and thus did not notice it; whereas for Pauling it provided additional evidence of the correctness of his model. This episode showed both Pauling's competitive spirit and his self-discipline. Finally, Pauling was sure enough of himself and of his model that he went ahead and published his findings on the alpha helix without having yet resolved the remaining discrepancy between his model and the available experimental evidence. First Pauling and Corey published a short note, followed by a longer article,[6] and then seven more papers to elaborate on their findings.

Pauling was a master at creating publicity for his discoveries. When he prepared to announce the discovery of the alpha helix, it was to be in a big lecture hall at Caltech. The model was on the rostrum covered with a piece of cloth, just like a statue waiting for unveiling. The moment everybody was waiting for came toward the end of Pauling's carefully choreographed lecture. It was like a crescendo in a musical movement, and when the model was finally unveiled, the audience was stunned by its beauty.

I experienced the mesmerizing effect of Pauling's lecturing style in 1982 at the University of Oslo. He covered the board with complicated formulas and from time to time looked at the audience as if checking whether we were duly impressed. Otherwise, the formulas were not at all necessary for us to understand the points he was making. He was already an octogenarian by this time, but watching him gave the impression of a young assistant professor who came for an interview and was presenting his research with the usual energy of such a scenario. During the lunch following the lecture, he was the most vigorous among us, leading the discussion and firing away questions, mostly answering them himself.

Pauling felt the need to communicate some details of the story of the discovery, and he wrote about them separately, but his article wasn't published until two years after his death. At that time, I was running a popular chemistry magazine, and his former secretary of his last twenty years, Dorothy Bruce Munro, suggested I publish it. Research papers usually lack the human element and the blind alleys followed in research, so this paper by Pauling is especially valuable to our understanding of how this particular discovery happened and how Pauling handled such a research project.[7]

The Cambridge group suffered a conspicuous defeat in this case—a defeat that was especially heavy for W. Lawrence Bragg to bear, because he was the pioneer of X-ray crystallography, and yet the American group had come out on top in this undeclared race. It was not possible to pinpoint a single reason for this defeat, but the fact that Pauling could restrict the circle of possible models because of his superior knowledge of structural chemistry was a critical detail. Years after this fiasco, Perutz complained about his group's lack of knowledge of the planarity of the peptide group. He blamed the Medical Research Council (MRC) for denying him the use of a Rockefeller Fellowship to travel to United States in 1948. The secretary of the MRC had thought that rather than going to learn from the Americans, the Americans should come and learn from the British. In hindsight, Perutz thought that he could have learned about the peptide bond planarity from Pauling, had he been allowed to travel.[8]

However, Perutz could have just walked across the street in Cambridge and asked the eminent chemist Lord Todd, who surely could have given him this information. Todd later remembered that he had mentioned to Perutz the double bond character of the peptide bond, but not being versed in the theory of resonance, Perutz did not make the connection to the planarity of the structure. It is not at all certain that if Perutz had visited Pauling he would have learned as much as he might have supposed in retrospect. We have seen Pauling withhold from Perutz his observation of the evidence of helical structure in Perutz's X-ray diagram. There is further evidence that Pauling was conscious of the competition. During his Oxford sojourn, he wrote to Corey in Pasadena that he was "beginning to feel a bit uncomfortable about the English competition."[9] Corey shared Pauling's sentiments when he remarked that he was "terribly impatient about getting into the protein work with a force that will really give the British some competition."[10]

The British considered protein crystallography their own territory, not only because the Braggs discovered X-ray crystallography and Astbury was a pioneer in taking X-ray pictures of proteins, but also because the Briton J. Desmond Bernal prepared for the first time X-ray diffraction diagrams of a protein—a pepsin single crystal—that clearly showed the possibility of deducing atomic positions from it. This was in 1934. As British crystallographer Dorothy Hodgkin describes it, "[T]hat night, Bernal, full of excitement, wandered about the streets of Cambridge, thinking of the future and how much it might be possible to know about the structure of proteins if the photographs he had just taken could be interpreted in every detail."[11] The British self-confidence in dominating this field reached such a proportion that Astbury and Bernal divided the structural research of biopolymers through a gentlemen's agreement between the two of them. They decided that Bernal would take up the investigation of the crystalline substances and Astbury, the fibrous ones.[12]

Perutz blamed Astbury's X-ray diffraction picture as well, which showed the same somewhat misleading repeat of the 5.1 angstroms that caused so much difficulty for Pauling. The difference was that Pauling eventually disregarded this evidence in light of the rest of his structural chemistry, whereas Perutz had nothing else on which to rely. Perutz was disheartened when he read Pauling's paper about the alpha helix model. He performed an additional X-ray experiment that gave further evidence, showing the correctness of Pauling's result, something that Pauling himself had missed.

When Perutz reported his finding to Bragg, Bragg asked him how he had thought of that, to which Perutz responded that it was because he was so angry that he hadn't thought of the structure himself. To this, Bragg replied coldly, "I wish I'd made you angry earlier."[13] Perutz told me this story in 1997, and he used Bragg's phrase as the title of his next book. Perutz thought that Pauling would be pleased that he, Perutz, provided additional evidence for the alpha helix model, but he was disappointed by Pauling's reaction. In Perutz's words, Pauling "could not bear the idea that someone else had thought for the α-helix of which he had not thought himself."[14] According to another British author, "Pauling's style was that of a conquistador in the realm of science."[15]

We add here a footnote about Pauling's theory of chemical resonance, which served him so well in the above story. At about the same time of his quest for the protein structure, this theory was in the center of attack by rabid ideologists in the Soviet Union.[16] The culmination was a four-day conference in Moscow in 1951, at the height of Stalin's tyranny in the post-WWII Soviet Union. The meeting was organized by the Soviet Academy of Sciences, and it was attended by leading Soviet chemists, physicists, philosophers, and others; even a poet attended the meeting. A small but vocal group of chemists attacked the theory of resonance as an ideological aberration; they also attacked quantum theory and the science of the West. They insisted on returning to traditional Russian values and advanced their own worthless theories. Excellent scientists suffered ruthless criticism for having applied the theory of resonance in their work, and they, in turn, offered humiliating self-criticism. The minutes of the meeting were published in a hefty volume.[17]

The affair has been referred to as the great Soviet resonance controversy, and it was a chapter in the antiscience events following World War II that touched biology even more severely. Physics was spared in the last minute, thanks to its decisive role in producing nuclear weapons. Stalin's terrorism did everything to protect his empire from even the slightest influence by the West, the purest sciences included. There is some irony in this story in that Pauling was a friend of the Soviet Union and suffered persecution in the United States during the McCarthy era: he was not allowed to travel abroad. But his friendly political views were not yet known in the Soviet Union. In 1993, I asked Pauling for his comments about this affair. He appeared as if he misunderstood the situation or did not want to understand it. He wrote me that it took years "for the chemists in the Soviet Union to get a proper understanding of the resonance theory."[18] In reality, they understood it well enough and applied it with great success; that is, until 1951, when the main proponents of the theory lost their jobs. They fared better, though, than some of their biologist colleagues who lost their lives in a similar ideological controversy.

As for Linus Pauling, he thrived on competition and remained active to the end of his life; perhaps his keeping himself busy in the competitive spirit prolonged his life.

Figure 5.1. Frederick Sanger in 1958, at the time of his first
Nobel Prize. Courtesy of the Medical Research Council Laboratory
of Molecular Biology Archives, Cambridge, United Kingdom.

CHAPTER 5

VALUING A STEADY JOB
Sequencing Biopolymers

[Sanger] is one of the great tool-builders of our time.
Freeman J. Dyson[1]

Frederick Sanger (1918–) has earned two Nobel Prizes in Chemistry. When he was asked about the main benefit from his awards, he said it was a steady job and good facilities for work. His response was characteristic of his personality. He never set the highest goals for himself; he just went about his work and solved the problems before him, step by step. His work made the Human Genome Project possible. After receiving his two Nobel Prizes, he continued his quiet life and focused on his work in the laboratory. When retirement age came, he retired just as quietly and has since devoted himself to gardening.

F rederick Sanger does not project the personality you might expect from one of the world's most distinguished scientists—and the only individual who has ever earned two Nobel Prizes in Chemistry. From a distance, it is easy to underestimate Sanger, and from his demeanor it is hard to discern his drive. Yet closer inspection suggests that although Sanger does not appear to be running, everybody else has to in order to keep pace with him.

He was expected to follow his father in his profession as a medical practitioner. But his father's work did not appeal much to him; visiting one patient after another seemed to him "a very scrappy sort of life."[2] There was a part of his father's work, though, that did entice him. The elder Sanger did some work in immunology: he identified bloods through immunological techniques, and even Scotland Yard became interested in the approach he had worked out.

Sanger realized that he would prefer working on a steady problem rather than dealing with patients, so he decided to study science instead of medicine. He admired his father for being useful to others, but he thought that as a scientist he could bring even more benefit to people. He had no clear idea, however, about what he was going to do. He knew that doing research was very prestigious, and he expected that it would be difficult to land such a job.

Sanger's brother, only a few years older, was the first important influence on him. Though they were very different, Sanger saw his brother as always the leader and just liked to follow him around. His brother's interest in nature and particularly in snakes and birds' nests rubbed off on Sanger early on. In school, Sanger did much better than his brother, who became a farmer after studying agriculture at Cambridge University.

Still, Sanger did not excel in his high school studies, nor did he win any scholarships, but he was fortunate enough to have parents who could afford to put him through higher education. When he arrived at Cambridge and had to choose his subjects, it was the first time he had heard of biochemistry. Eventually it became his favorite subject because he found it exciting that biology could be explained by chemistry, and it was taught by a highly enthusiastic instructor by the name of Ernest Baldwin.

Sanger majored in chemistry and put in an extra year in 1939 to take an advanced course in biochemistry. That course was supervised by a refugee scientist from Germany, E. Friedmann, who stressed the importance of isolation and structure in biochemistry.

When Sanger took his final examination, he was somewhat unsure of himself, and so was surprised to learn that he had earned a first-class degree. He received his BA in 1939. There were plenty of research positions at Cambridge at the time, and since he was not called upon to do military service, Sanger was able to continue his studies for his doctorate without interruption.

Brought up as a Quaker, Sanger became a conscientious objector. He signed the Peace Pledge Union, which was started by a Church of England priest. Sanger became a Quaker because his father had joined the Quakers when he was thirty years old and had become keenly observant. Sanger, however, gradually became less interested in the religion, and had there been another call to arms later, he was not sure what he would have done. His studies of science brought about a change in his outlook on the world. He found it increasingly difficult to believe something that he did not know, and in science he felt he had to be careful about what he considered as "truth." He "found that it was difficult to believe all the things associated with religion."[3] At one point, he started calling himself an agnostic. "If you are a scientist and believe in truth, you've got to say you don't know when you don't know. . . . There's always a temptation, especially if you have a theory, to try to prove it rather than to find out what is the truth."[4]

Sanger started his doctoral studies with N. W. (Bill) Pirie, who was involved with protein chemistry and whose topic at the time—appropriate, considering the war conditions—was making edible protein from grass. Pirie soon left for another job, and Albert Neuberger, a postdoctoral fellow, became Sanger's doctoral supervisor. Neuberger's main interest was protein metabolism; that is, he studied what happens to proteins in the living organism. Sanger considered Neuberger his main teacher. He taught him "both by instruction and by example" how to do research.[5] The title of Sanger's thesis was "Metabolism of Lysine," lysine being one of

the natural amino acids. According to Sanger, with his characteristic modesty, his doctoral work was "nothing very profound, but
through it I gained much experience in amino acid chemistry, a
good introduction to 'sequences.'"[6] Sequence is the order in which
building blocks are linked to each other in macromolecules; thus,
for example, it is the order of amino acid units building up a protein molecule.

Sanger had to learn a lot about proteins in the course of his doctoral work. Proteins consist of long chains of amino acid units, as
was first established by German chemist Emil Fischer around the
turn of the nineteenth century. At the time Sanger was entering the
field, there were still controversial views about the composition of
proteins, but the Cambridge biochemists followed Fischer's lead.
There were pioneers of protein chemistry in Britain who further
developed the field and ascribed well-defined composition to proteins. They believed that the protein chains had structures that
could be determined. The X-ray crystallographers had already
taken some diffraction photographs of biological macromolecules,
and soon some rudimentary models of protein structures appeared
(see chapter 4).

Sanger was to spend all his life in Cambridge and its vicinity.
He received excellent training, which included plenty of opportunities to connect with the pioneers of his field and related areas.
There, he could often learn about the latest methodological innovations in chemistry. These innovations included excellent new tools
of analytical chemistry that made it possible to separate and identify minute amounts of substances.[7] Sanger was pushed by future
Nobel laureate Christian B. Anfinsen, who visited him in Cambridge in 1954, toward using radioactive isotopes. Illustrious visitors went on pilgrimages to Cambridge because they knew they
could learn something new there, in addition to having the opportunity to inform others at one of the best venues in science about

their own recent achievements. Anfinsen, for example, benefited from observing Sanger's work and learned about the connection between sequence and biological activity.

Soon after Sanger earned his doctorate, Neuberger left the biochemistry department. Sanger, however, was very lucky—again—because he got a job right away at the department. At this time, Albert C. Chibnall had just been named the second Sir William Dunn Professor of Biochemistry, and Sanger became his associate. Chibnall took an interest in Sanger's research and suggested that he investigate the protein insulin. He recommended starting with the determination of the end groups of its chain and setting as a next goal the determination of the number of amino acids in proteins. This was a step in the direction of determining the sequence of amino acids in proteins, an approach nobody had tried before.

Chibnall was an outstanding plant biochemist and a modest person who did not feel quite comfortable in his prestigious chair. He thought that a medically qualified biochemist should occupy the position and resigned in 1949. He stayed on at the department, continued his protein research, and supported Sanger's work on insulin, in which he saw great promise.

Insulin was already known as an important protein, but Chibnall's choice was motivated by the fact that it was the only protein at the time that could be purchased in a pure form. Insulin is a rather small protein as far as proteins go; it consists of fifty-one amino acids. Sanger did not set out to determine the sequence of these amino acids, only the end groups. In fact, he did not think about the complete sequence until the final stage of the work when he was identifying the various groups in insulin. This was fortunate, because at the very end of the work, he started feeling the extra pressure generated by the expectation of producing the complete sequence. It had already taken several years to reach that point.

In the course of his work on insulin, Sanger in 1947 received an invitation to spend a few months as a guest researcher in Sweden at Arne Tiselius's laboratory at the University of Uppsala. This was a unique opportunity for Sanger, who, up to this point, had not had much experience outside of Cambridge. Tiselius was a great authority, Uppsala had first-rate instrumentation, and Sweden was a land of plenty at this time so soon after the war; life in Britain was still suffering from shortages. Tiselius was dedicated to the study of complex protein systems, and he invented popular techniques for their investigation. For his research, he was to receive the Nobel Prize in Chemistry in the following year, 1948. He became an influential force in the Nobel Prize institution for many years to come.

Sanger had a mixed experience at Uppsala. Probably due to a technician's error, some data suggested that insulin contained four chains (rather than two, as was soon determined). When Tiselius learned about this seemingly new result, he suggested to Sanger that he send a letter to *Nature* under their two names. The careful Sanger was rather taken aback by this haste and by Tiselius's intention to be named coauthor without having contributed anything to the work.[8] Sanger now understood—more than he had before—the advantage of working under Chibnall, who had helped him selflessly. In addition to the unpleasant undercurrent from this experience with Tiselius, the conclusion of the report turned out to be wrong. It would be the only publication Sanger ever felt uncomfortable about.

Soon it was determined unambiguously that insulin contained two chains: one consisting of thirty amino acids, the other of twenty-one amino acids, and the two chains were linked by sulfur–sulfur bridges. Sanger first determined the end groups of the chains, then the next groups, and thus shorter and shorter segments remained unknown. He then broke down the long chains into smaller fragments, isolated and identified them, and gradually— indeed, very gradually, after years of painstaking work—he succeeded in determining the complete sequence of insulin.

Sanger had excellent associates in his quest for the insulin sequence. His first doctoral student, Rodney Porter, was slightly older than him. Porter had fought in World War II, so he was behind with his studies. When Porter joined Sanger, Sanger's techniques could already be applied to proteins other than insulin. This was fortunate for Porter, because he developed an interest in the structure of another protein, gamma-globulin. He worked in a timely fashion and received his PhD degree in 1948. He went on to a shining career crowned with a chair of biochemistry at the University of Oxford, followed by the Nobel Prize in 1972 for his discoveries in the chemical structures of antibodies.[9]

Two of Sanger's associates were involved in sequencing insulin. Hans Tuppy from Austria spent a postdoctoral year with Sanger and worked out most of the sequence of the longer chain. E. O. P. (Ted) Thompson from Australia worked on the shorter chain, which was no easier. In addition, Sanger and Thompson had to determine the locations of the sulfur bridges connecting the two chains, and this was another challenge. Both Tuppy and Thompson went on to distinguished careers. Thompson remained in science; Tuppy became a preeminent administrator in Austrian science.

The completion of the insulin sequence was a great achievement, and its significance by far outgrew insulin alone. The techniques Sanger used for insulin proved to have general validity and could be applied to other proteins as well. Sanger thus established a technique for determining the sequence of amino acids in proteins. This was a discovery, but it was a different kind of discovery than typically imagined by the public.

Discoveries often take place over a very short period of time, and "Eureka" moments can often be identified, even if preceded by a great deal of painstaking effort. Sanger's discovery was a long process, advancing step by step. No single big "Eureka" moment could be pinpointed at the end, though many smaller such moments took place during the process. Sanger's discovery had a tremendous impact on biochemistry and on the biomedical sciences. He was awarded an unshared Nobel Prize in Chemistry for it in 1958. At the award ceremony, Tiselius was the Swedish academician who gave the traditional presentation speech.

Sanger eloquently and without undue modesty spoke at the conclusion of his Nobel lecture about his expectations of the impact his technique would have: "The determination of the structure of insulin clearly opens up the way to similar studies on other proteins and already such studies are going on in a number of laboratories. These studies are aimed at determining the exact chemical structure of the many proteins that go to make up living matter and hence at understanding how these proteins perform their specific functions on which the processes of Life depend."[10] Sanger then—as if referring back to his youthful decision to go into science rather than medicine—expressed his hope that studies on proteins would reveal changes taking place in disease, and hence their efforts would provide practical use to humanity.

Nobel laureates are usually bogged down for a year or so by the celebrations and by media interest in their work, life, and experience. Eventually they have to decide in which direction they should continue their lives. Sanger was forty years old at the time of his first Nobel award, and he abhorred the thought of taking up a high administrative leadership position. For a short while he considered teaching, but he realized that he was not especially good at lecturing. Actually, at one point, he tried to improve his lecturing style and carefully inserted jokes into his presentation, à la Francis Crick, but it did not work for him; the jokes came out flat and were received in silence.

He soon understood that of all his options the only one he truly enjoyed was doing research—what he had been doing before his Nobel Prize. One of the benefits from his new fame was that it became yet easier for him to attract good students and particularly well-trained postdoctoral fellows. So Sanger continued to study proteins, but he had not yet planned to sequence nucleic acids. He thought that the nucleic acids would be too big for such experiments, and that they would be especially difficult to sequence because they consisted of only four different components (called nucleotides). Sequencing a very long molecule with only four components would be much more difficult than working on shorter molecules with a much greater diversity of components, as was the case with proteins.

One of Sanger's strengths in research was his affinity for new techniques. His new analytical approaches were important because he was always facing the task of identifying products of the reactions he carried out. Often, he learned about the emerging techniques through interactions with others at meetings or from visitors. Thus he did not have to wait for the descriptions of new techniques to be published, and he was an early user of the latest innovations. He himself found great fun in developing new methods, but once he had developed them, he found no challenge in their repeated applications. Once a new sequencing method had become routine, he tried to avoid it because he did not feel comfortable being one of the many researchers who performed repetitive work in which his innovative approach to problems had no outlet.

It is also of interest to see how Sanger handled those periods when success eluded him. He called them "the lean years."[11] Such periods occurred both before and after he was a Nobel laureate. He did not like to dwell on an unsuccessful experiment; rather, he preferred to start planning the next experiment right away. He realized that this attitude resulted in fewer published works; still, he felt fortunate that he could afford going for extended periods without having to write papers. He had a permanent position with the Medical Research Council (MRC), and he was able to work on long-range projects at his own pace, investigating challenging problems that did not promise immediate results. There was one drawback to his approach: such style of work could be rather disadvantageous for the careers of Sanger's young associates.

When the MRC opened its Laboratory of Molecular Biology (LMB) outside of downtown Cambridge, Sanger left the Biochemistry Department and joined Max Perutz, Francis Crick, and others in the new facilities. He enjoyed the plentitude of space, but when the laboratory became crowded again, he saw advantage in this, too, as crowded conditions are conducive to fruitful interactions. With a limited number of openings, Sanger found it difficult to choose among the many excellent students and postdoctoral fellows whose applications flooded his office. The letters of recommendation often failed to convey the most important traits of the applicant. In the past, Sanger had looked for associates who complemented him by being more outgoing than he was. Eventually he enjoyed having people around him who were similarly quiet and reserved, and he saw this change in terms of his own growing maturity.

Sanger had regularly attended the Gordon Research Conferences in New England—informal meetings covering many areas of science named after the initiator of the conferences. Although the meetings on proteins and nucleic acids were combined, Sanger preferred attending the protein lectures because he found the talks about nucleic acids boring. Gradually, however, some of those talks about nucleic acids rubbed off on Sanger, and eventually his interest turned toward nucleic acids. Nucleic acids consist of phosphates, sugars, and nitrogenous bases; the phosphates and sugars repeat uniformly in them, and five different bases participate in them, building up the two principal kinds of nucleic acid: deoxyribonucleic acid (DNA) and ribonucleic acid (RNA). Sanger initially considered DNA impossible to sequence, but he was not so sure about RNA. The so-called transfer-RNA, whose function was transferring genetic information, was relatively short; it contained about seventy or eighty nucleotides. Finally, around 1965, Sanger started working on RNA.

In the meantime, another scientist, Robert Holley, and his colleagues had performed the first successful sequencing of another RNA molecule. They used the method that had been worked out for proteins. Holley's success probably stimulated Sanger, who was already set on moving into nucleic acid sequencing. Sanger worked out a technique for separating small degradation products of RNA, and together with his PhD student, George Brownlee, used it to determine the sequence of an RNA molecule with one hundred and twenty nucleotides. This was the largest system to date (around 1965). In the course of this work, Sanger developed new techniques for various steps in sequencing.

Although Sanger's years of painstaking work on this project did not—again—culminate in one big "Eureka" moment, there were many joyful steps along the way. One such moment occurred when, after many trials resulting in uncertain and foggy patterns, Sanger's group produced something that was clear and unambiguous. Outsiders might not have understood the significance of this, but the team knew when yet another big step forward had been made.

Sanger had many more coworkers on the nucleic acid project than he'd had on the protein project. Gifted American postdocs flocked to his laboratory, eager to learn from him and be part of this exciting adventure. His closest collaborators were two British assistants, B. G. (Bart) Barrell and Alan Coulson. What Barrell lacked in academic qualifications, he made up for with knowledge and enthusiasm. Coulson, on the other hand, was rather withdrawn, like Sanger. Barrell found much joy in actual sequencing, whereas Sanger continued to be fascinated in finding new methods—a pursuit he shared with Coulson. But these were only two of the numerous associates Sanger worked with during the years of nucleic acid sequencing.

The climax of Sanger's work was a new method that made it possible to sequence large DNA molecules. The dideoxy method—the name of which came from a set of important substances utilized in the technique, the dideoxynucleotide triphosphates—could be used directly with large molecules. The smallest DNA molecule Sanger could acquire to work with was the DNA of a bacteriophage, which contains about five thousand nucleotides. The approaches he and his team had used previously (based on degradation) did not work satisfactorily, so they started looking for new ones. This was when they began using the DNA polymerase enzyme, which normally synthesizes DNA chains. Finally, they made use of a similar substance, dideoxynucleotide triphosphate, which incorporated itself into the growing DNA chains but acted as terminator to further extension. If the same dideoxy derivatives were used for all chains, all chains would end at that particular nucleotide. In other words, they could break up longer chains into smaller ones having the same reference points at one of the ends of these smaller chains. This approach was successfully utilized in sequence determination, and was augmented by devising a very efficient separation technique, which sorted out the nucleotide chains exactly according to size. This was called *polyacrylamide gel electrophoresis*. The combination of these two techniques made it possible for Sanger and his team to sequence DNA fragments of almost any size in a rapid and simple way. Sanger considered this method the culmination of his research career.[12]

At about the time the dideoxy method for sequencing large DNA molecules was completed, another method had been worked out at Harvard University by Walter Gilbert and Allen Maxam. Sanger's method involved enzymes, whereas the Maxam–Gilbert method operated through more chemical approaches. Sanger was disappointed to learn of a competing method after so many years of painstaking work, but he recognized that the two methods were to some extent complementary. He thought that the Maxam–Gilbert method was more convenient to use for short, specific sequences, whereas the dideoxy method was preferable for long sequences.

Sanger compiled a table of milestones in sequencing research, starting with some preliminary work as early as the mid-1930s, and ending in the mid-1980s with an entry of a virus containing 172,282 nucleotides. Next came the sequencing of organisms of ever-increasing size and the complete sequencing of the human genome, but those accomplishments belonged to others—yet they all had their roots in Sanger's work.

The Nobel Committee of Chemistry recognized the importance of the methods for sequencing nucleic acids and awarded half of the 1980 Nobel Prize to Walter Gilbert and Frederick Sanger "for their contributions concerning the determination of base sequences in nucleic acids." At this stage of his career, Sanger required no fur-

Figure 5.2. Walter Gilbert and Frederick Sanger at the Nobel Centennial in Stockholm, 2001. Photo by the author.

ther fame or recognition, but Gilbert was due the acknowledgment. At any rate, he would not have been recognized without Sanger, hence Sanger's second Nobel Prize.

Sanger and Gilbert have different professional backgrounds; Sanger's is in biochemistry, and Gilbert's in theoretical physics. They also have very different personalities. For example, Gilbert has a flair for entrepreneurship, whereas Sanger never got involved in patenting, let alone in founding and directing companies.

Sanger declined knighthood because he did not want to be distinguished from everybody else by being addressed as "Sir Frederick." He was always "Fred" to everybody at the Laboratory of Molecular Biology. However, Freeman Dyson, the distinguished British-American physicist and author, assumed that Sanger was "Sir Fred Sanger," and also that he was the director of a large enterprise.[13] He was neither.

Elizabeth Blackburn, just out of Australia, was Sanger's doctoral student in the early 1970s. She felt comfortable in the environment of Sanger's unassuming, low-key laboratory. Her biographer noted that "Sanger imparted a curiosity-driven, pragmatic approach to research: 'just get in and find the sequence.'" To this, Blackburn added, ". . . the courage to just wade in and try, if you thought it was a good idea, came from Fred's general philosophy."[14] Blackburn went on to a brilliant carrier in biology and was corecipient of the Nobel Prize in Physiology or Medicine in 2009.

When Sanger says that the main benefit from his Nobel Prizes was that he had a secure job and good facilities for research, one has to believe that he is being genuine, although such a statement coming from someone else might not sound plausible. Those who have known Sanger from the time before he received his first Nobel Prize vouch for his having remained the same person over the decades. Modesty is perhaps the most characteristic trait one might assign to him, yet he is also a very sophisticated person whose

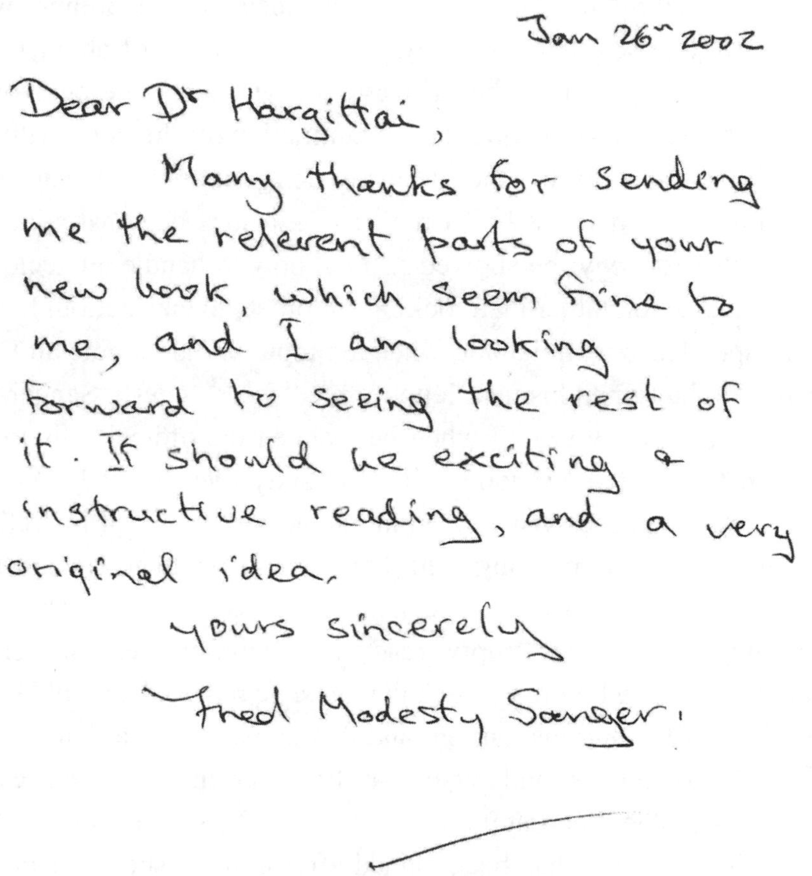

Jan 26ᵗʰ 2002

Dear Dr Hargittai,

Many thanks for sending me the relevant parts of your new book, which seem fine to me, and I am looking forward to seeing the rest of it. It should be exciting & instructive reading, and a very original idea.

yours sincerely

Fred Modesty Sanger.

Figure 5.3. Letter from "Fred Modesty Sanger" to the author, 2002.

modesty might be more than his natural self-expression. I mention here one example.

Molecular biologist Sidney Altman shared with me a story from the days long before he received his own Nobel Prize, when he was doing his postdoctoral research at the MRC LMB. The bubbler tube he was using in an experiment became punctured, and radioactive material scattered all over his work area. He looked for the radioactive safety officer, but the officer was not at the lab. Next, Altman told lab head Sydney Brenner about the accident, but Brenner was too busy and sent Altman away. When Sanger heard about what happened, he put on rubber gloves and, with a sponge and some detergent, started collecting the contamination off the floor. Altman considered this to be the ultimate demonstration of Sanger's humility and greatness.[15] To me, it was also a demonstration of Sanger's pedagogy: he showed Altman how to handle an accident by taking action rather than looking for others to take action. In my correspondence with Sanger, when the topic of his humble attitude came up, he signed his next letter to me "Fred Modesty Sanger."[16]

Sanger quietly retired when he reached the official retirement age in the United Kingdom—he was sixty-five in 1983. One of Sanger's colleagues visited him in his lab on the eve of his retirement and found him sitting at his bench in his white gown carrying out an experiment as if he were in the middle of his career. The next morning, his lab was empty, ready to receive its next occupant. Sanger had long before decided that once he retired, he would busy himself by tending his garden, and this is exactly what happened. Apart from occasional visits—only when he is specifically invited—to the Laboratory of Molecular Biology or to the big genome center that has been named after him, he spends his days gardening.

CHAPTER 6

SAVING TIME AND LABOR
Combinatorial Chemistry

Nowadays combinatorial chemistry is an accepted branch of chemical science. Fifteen years ago it was completely unknown.
Árpád Furka in 1999[1]

Árpád Furka (1931–) wondered about the great variety of peptide compositions that would take many generations of scientists to synthesize. Eventually, he worked out a simple technique that made it possible to enhance the efficiency of such work by a million or even a billion times. From this, a new field of chemistry emerged, called combinatorial chemistry. Furka came from such a humble background that he was an unlikely candidate to become a scientist, let alone a discoverer. He overcame the hurdles of his roots to eventually achieve universal recognition among his peers.

Protein synthesis was a major achievement in twentieth-century science. Part of the process was a quest to find out how nature synthesizes proteins by following the genetic information contained in nucleic acids: when the genetic code was cracked, the information transfer from nucleic acids toward proteins was understood. The other area of investigation of protein synthesis was the production of proteins in the laboratory from their building blocks, the amino acids. Proteins are very long chains of amino acids,

123

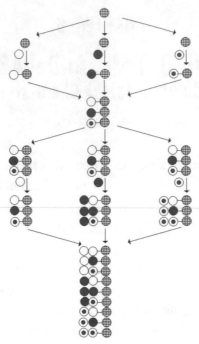

Figure 6.1. Illustration of the initial stages of the application of the combinatorial principle for peptide synthesis. Prepared by and courtesy of Árpád Furka, Budapest.

whereas peptides are just short chains of amino acids. There is no strict division between the peptides and proteins. The great German chemist Emil Fischer synthesized the first peptide around the turn of the nineteenth century. The number of naturally occurring amino acids building up proteins is twenty, and this was established in the middle of the twentieth century.

The synthesis of a protein in the laboratory starts with one amino acid. Another is linked to it, and so on. The linkage between two amino acids is the peptide bond. (We have described this peptide bond in our discussion of Linus Pauling's quest for the structure of proteins in chapter 4.) Even the synthesis of the short peptides used to be a time-consuming procedure until biochemist

Bruce Merrifield introduced a new approach called synthesis on a solid matrix. Merrifield worked at Rockefeller Institute for Medical Research (today, Rockefeller University) in New York City, and his supervisor, D. Wayne Woolley, gave him the task of preparing a simple peptide of five amino acids.[2] It took Merrifield several months before he could produce this substance, even though he was diligent and knowledgeable about the existing techniques.

The experience prompted him to think that there must be a better way, and one day in 1959, he had the idea of anchoring the first amino acid of the chain onto a solid support, something that would not dissolve in any of the solutions used in the process. This is how he started the synthesis. He then added the second amino acid to the first, and the third, and continued building up the chain. After each addition of the next amino acid, he would wash away the substances that were no longer needed while his target peptide stayed attached to the insoluble support. Thus he had done away with the previously necessary purification and crystallization of each intermediate, which had been the standard way to proceed. When the desired length of the peptide was reached, he broke the bond to the solid support and had his peptide ready.

It was a simple procedure, but it took Merrifield years to develop. He finally published his technique in 1963,[3] at which time the Rockefeller Institute was reorganized into Rockefeller University. Merrifield owed a lot to the ideal conditions for research at Rockefeller. His superior, Woolley, was also exceptional; in spite of being blind, he was an outstanding chemist, and selfless, too. He let Merrifield work on his idea for a long time without publishable results, and when Merrifield was ready, Woolley encouraged him to publish his invention alone, lest his achievement be attributed to his better-known boss. Merrifield's method revolutionized peptide synthesis, and in 1984 he was awarded the Nobel Prize in Chemistry for it.

It was not long after Merrifield's new process of peptide syn-

Figure 6.2. Bruce Merrifield at Rockefeller University in New York City, 1996. Photo by the author.

thesis had become known that Hungarian chemist Árpád Furka started thinking about peptide synthesis, but from the opposite end of considerations. He was doing postdoctoral work at the Department of Biochemistry of the University of Alberta in Canada in 1964–65, where his task was to determine the sequence of the amino acids that made the digestive enzyme called chymotrypsinogen B. His host, Professor L. B. Smillie, had become acquainted with the technique of sequencing in Cambridge, United Kingdom, where Frederick Sanger had worked it out (chapter 5). Furka had a formidable project, because this protein comprised 245 amino acid units, and any of the 245 positions in the chain could be occupied by any of the twenty naturally occurring amino acids. He solved the problem almost completely by the time his postdoctoral year was over and then returned to Hungary. However, he did not stop thinking about the possible variations of the composition of protein chains.

Furka wanted to get an idea of how many different sequences might be possible for a protein having 245 units. He was astonished by the magnitude of the number at which he arrived: 20 raised to the power of 245 (in scientific notation, 20^{245}), where 20 is the number of possible different amino acids at each of the amino acid positions in the chain, with a total of 245 amino acids forming the chain. The number 20^{245} is so huge that we cannot possibly imagine its magnitude. In more common scientific notation, it could be written as 5.65×10^{318} or 565×10^{316}, which could also be written as 565 followed by *316 zeros*! Even if only one single molecule were prepared from each of these variations, there would not be enough matter in the whole universe to form them. Furka wanted to get a grasp of the number of variations of proteins, but he could not fathom such numbers, so he turned to the much smaller peptides.

To calculate the number of variations for the small peptides comprising a few amino acid units, Furka had to use the same method: 20 raised to the power of the number of amino acid units in the peptide. At the beginning, the numbers were much smaller, though they quickly increased. A dipeptide has two amino acids, and since each of the two may be any of the twenty different amino acids, the total number of variations is $20^2 = 20 \times 20 = 400$. A tripeptide has three amino acids, and the number of possibilities is, accordingly, $20^3 = 20 \times 20 \times 20 = 8,000$, and so on, as given here,

Number of amino acids in the peptide	Number of possible sequences
Dipeptide, 2	400
Tripeptide, 3	8,000
Tetrapeptide, 4	160,000
Pentapeptide, 5	3,200,000
Hexapeptide, 6	64,000,000
Heptapeptide, 7	1,280,000,000
and so on.	

These numbers were less frightening than the ones Furka had obtained for his protein, but they quickly grew improbably larger as the number of amino acid units kept growing. When Furka considered the amount of time and labor that would be necessary to produce all the variations of these peptides, he was awestruck by the results.

It might be asked whether Furka's question made any sense at all. Would there be any point to producing all the possible variations of a peptide chain, let alone a protein chain? In principle, yes, because it could not be excluded that among the many variations some might be useful for biomedical purposes. It is a brute-force approach to produce all possible compounds and to look for those that might be useful, but this is a possible route in drug design. In addition to producing the compounds, they would have to be screened for biological activity against all possible targets, which would take a lot of additional time and labor but would also carry the promise of finding something beneficial.

Furka estimated that even if he used Merrifield's technique, it would take a day to add another amino acid to the chain, and at that rate, the preparation of a pentapeptide—not a very large molecule—would take five days. The number of variations of pentapeptides is 3.2 million; accordingly, it would take 3.2 million days; that is, 43,800 years of uninterrupted work for one scientist to produce all variations. Furka took into account all possibilities that might help reduce the amount of time, but he still could not reduce it to less than thousands of years. He continued thinking about this problem because he found the accessibility of all peptide sequences important for drug design.

He came up with various ideas and discarded most, but around 1980, one idea stuck. It will be outlined here in a simplified manner. Furka relied on Merrifield's methodology, but he divided the solid support into twenty equal portions, attached a different amino acid to each portion, mixed the portions, and continued dividing, adding, and mixing, and, again, dividing, adding, and mixing, and so on. He repeated these cycles until he achieved the desired length of the peptides, as shown in Merrifield's methodology. But there was a difference: in Furka's procedure, *all the possible combinations* of the amino acids present in the reaction mixture were being formed, and this is why we call it a *combinatorial procedure*. This procedure yielded *simultaneously* all possible variations of the amino acid sequences for a peptide of given length. Furka's portioning-and-mixing approach for *all* variations of the peptide of a given length did not take any longer than Merrifield's method to produce a *single* peptide of the same length.

By 1982, Furka had a description of his combinatorial procedure, and it was time to look for applications. He was aware of the need for producing peptides of great diversity for pharmaceutical research, and he had already been involved with one of the largest pharmaceutical companies in Budapest. However, when he showed his ideas to the researchers at the company, they voiced no interest in his innovation—it was so revolutionary that they just did not have the right frame of mind for it. Furka thought that it might at least be worthwhile to patent his findings, so he brought his idea to the patent officer of the company. She did not think it was patentable, although some years later several other researchers would file for patents for this procedure in the United States.

The patent officer had reservations, but she wanted to help Furka, so she advised him to have the description of his methodology notarized. This he did on June 15, 1982, and it became the first document describing combinatorial chemistry. In 2002, it was published in *Drug Discovery Today*, and it can be seen on Furka's home page.[4] Furka's method has become known by the short name of the "split-mix method."

Árpád Furka was a latecomer to science, which was not surprising. What was surprising was that he came to science at all.[5] He

Figure 6.3. Árpád Furka at Eötvös University in Budapest, 1999. Photo by the author.

was born to Hungarian parents in Kristyor, Romania, a village in Transylvania. The village was located in a region of gold mines, and Furka's father worked for one of the mining companies. Furka had five siblings, and the family found it hard to make ends meet. They spoke Hungarian at home, but Furka attended a Romanian school until the age of eleven. His mother was originally from Hungary, and in 1942 she decided to return. She took the children with her, except for one of her daughters, who by then had married. Furka's father stayed in Romania because he did not want to forfeit his pension. For a few years, Furka continued his schooling in Hungary.

Furka started working when he was still a child. The family needed whatever he could contribute to ease their hardship. When he completed the mandatory general school at the age of fourteen, he did not go to secondary school but instead looked for employment. It was a difficult time: the war had just ended and the country was in ruins. Furka managed to find only odd jobs from time to time. Instead of receiving money for his labors, Furka was paid in kind. When he helped harvest wheat, for example, he earned enough wheat to help feed his family during the winter. He spent the next five years working in this manner, but he always regretted that he could not continue in school, and he read everything that came his way.

By 1950, Hungary had become a communist country, and the new regime needed its own recruits in positions of responsibility. Children who had not had the opportunity to study previously were encouraged to enroll. This new arrangement seemed tailor-made for Furka. Normal high school was four years, but new cadres were needed sooner. Also, the candidates for continuing education were older than the typical school-age children, so they were eager to get their education and move on with their lives. Accelerated courses were organized and offered for the would-be students in order to complete secondary education in one year instead of the customary four.

Furka was invited to take such a course. He found it tough.

Although students like him had skipped many of the subjects of the four-year curriculum that were deemed nonessential, they still had to learn much more than the typical schedule would have prescribed for one year. But what they had to learn, they learned well. They had a rigorous timetable, lived at the school, and could leave only on weekends. Their graduation picture shows thirty-eight sober-looking young men. All thirty-seven graduates fashioned open collars, with only the teacher wearing a tie. The certificate earned at the end of their studies was called something like a special matriculation, and the term acquired an ominous connotation: it signified a deficient education to some and tough perseverance to others. Those who received this education missed most of the cultural aspects of high school, but they regained some of the knowledge from their lost years.

At that time, the special matriculation gave students an advantage in being admitted to a university, where many of these graduates were sent to continue their education. Furka was not asked about what he would like to study; he was just assigned to train to become a high school teacher of chemistry and physics. He was directed to the University of Szeged in southern Hungary. (Albert Szent-Györgyi had worked there, but by this time he had moved away and was already in the United States.) Furka did not mind that he had not been asked about his preference; he was grateful for the opportunity, and he always liked the sciences. His dream had been to become an astronomer, which he realized would have not been practicable. Studying chemistry and physics seemed to be the closest to his dream.

Furka's knowledge from his speedy high school education combined with his thirst for learning successfully withstood the rigors of university studies. Not only did he keep up with his peers who had come to the university by the normal route, but he was often able to help them in their studies. He received his teacher's certificate in 1955 and was sent to teach in a small town, Makó, in southeastern Hungary. After one year, however, he was called back to the University of Szeged, where he was charged with teaching in the Department of Chemical Technology.

Because university instructors were supposed to get involved in research, Furka came under the influence of the professor of organic chemistry, Gábor Fodor. However, Furka's work with Fodor had barely started when the Hungarian Revolution broke out on October 23, 1956, followed by the Soviet Union's ruthless crushing of it. These events did not take more than a few weeks, but the ensuing general chaos and striking in the country lasted for months. Furka remained aloof from politics; he felt he had a lot of catching up to do, and he used the forced break for studying English. When conditions permitted, he was happy to resume his research, and he found an inspiring mentor in Fodor. Alas, their interactions did not last for long. Fodor soon left Szeged for Budapest, then immigrated to Canada, and finally ended up in the United States. He became a professor at the University of West Virginia and built up a strong research environment there.

Furka's life continued as before, and in 1961, soon after he had earned his PhD-equivalent degree, he was transferred from Szeged to Budapest, again, without having been asked about his intentions or preferences. He was told to join the Department of Organic Chemistry at Eötvös University, where he found the atmosphere to be quite chilly. In time he would learn that the department head, Professor Viktor Bruckner, had wanted to fill the opening with someone else, when Furka was essentially forced down his throat. Regardless, Furka did not experience overt hostility, and nobody hindered his work in any way.

Furka's alienation from the rest of the department was as much his own doing as that of his environment. Professor Bruckner was much revered in his field. A few years after Furka had come under his direction, I attended Bruckner's course in organic chemistry. He was an old-fashioned gentleman who held a weekly afternoon tea for his leading associates—the structure of the department was hierarchical with the professor at the top—and whether or not he liked Furka, he sent word to him that he would be welcome to join these afternoon tea gatherings. Furka had no background to understand that this was a social occasion having very little to do with who was drinking what. He took the invitation literally and sent back word to the professor that he did not like tea. This was impolite on more than one level, but Furka realized this only years later.

Furka's only joy at work was getting into the field of peptide chemistry. He started a notebook in which he recorded his thoughts, and when he found something he thought might be of interest, he took it to Professor Bruckner, with whom he had built up a working relationship. To Furka's regret, the professor never showed interest in any of the ideas presented to him. According to the old notebook, Furka scribbled down some thoughts about solid-state synthesis of peptides. This was in 1962, still before Merrifield's publication about his new method of peptide synthesis.

When Furka applied for a postdoctoral fellowship advertised by the National Research Council of Canada, he received backing from the authorities in what was probably their last interference in his life. As a result, he spent a fruitful year in Edmonton. Upon his return to Budapest in 1965, Furka embarked on working out new methodologies in protein and peptide chemistry that did not need expensive instrumentation. The next step in his advancement was earning a higher doctorate, the doctor of science degree. This made it possible for him to get a professorial appointment. The old Prussian-style system of one department/one professor in Hungary, too, was giving way to several professors for one department—and Furka was appointed full professor in 1972. The professorial appointment gave him complete independence in research, and using this new freedom, he entered a fruitful interaction with the Chinoin pharmaceutical company. He became involved in isolating natural peptides and in the determination of their amino acid sequences. His team synthesized these naturally occurring peptides and various analogs—whatever the industry people required in their search for new drugs. This work contributed to Furka's revolutionary ideas of combinatorial chemistry.

Once Furka understood that he could not have his methodology patented, he started thinking about publishing it. Simultaneously, he and his colleagues at the university—he was building up a small research group there—embarked on the experimental realization of his ideas. They used a modern technique invented recently in Cambridge, United Kingdom, to identify the emerging peptides on a sheet of paper. The technique utilized the differences in mobility of various peptides as determined by their charges and masses. Furka's group received a big boost when a doctoral student from Ethiopia, Mamo Asgedom, joined them. Asgedom's thesis was based on these very experiments.[6] Alas, Asgedom soon returned to Ethiopia, but Furka has always given proper credit to him whenever their joint work is mentioned.

Furka had talked with many of his organic chemistry colleagues about his ideas and methodologies before trying to publish them. Their strong disbelief, ridicule, and refusal even to enter a meaningful discussion discouraged Furka. It took a long time before he decided to come out into the public arena with his invention. Finally, after six years from the time he had attempted to patent his invention, he exhibited two posters at two international meetings in quick succession. His posters did not generate any particular interest among those who saw them, but at least his abstracts have become part of the record.[7]

Furka deliberated for a long time over where to publish an article on his invention. He thought that general-purpose magazines like *Nature* would not take his manuscript, so he decided on a more specialized venue where he had published previously: the *International Journal of Peptide and Protein Research*. Furka and his associates prepared a manuscript, which arrived at the editor's office in Tucson, Arizona, on February 12, 1990. Three reviewers read the manuscript, and the editor wrote to Furka on May 15, 1990, that the manuscript might be acceptable only after major revision. The group in Budapest carried out additional experiments and mailed the revised manuscript back to the journal, whereupon it was accepted for publication on November 21, 1990. It was published in June 1991.[8]

For many years, Furka's contribution to this field has been downplayed in science journals and elsewhere, and he often felt as if a wall of silence was surrounding him. Gradually, however, many in the field have become acquainted with his original publications and have started making references to them. Lately he has been increasingly referred to as the pioneer or the father of combinatorial chemistry. When the European Society of Combinatorial Sciences held its first symposium, in Budapest in 2001, Furka was elected honorary president of the organization.

I have known Furka for more than forty years and have known him as a person of integrity and inventiveness. He advanced original ideas before his combinatorial chemistry and since then, as well. I have talked with him quite a bit about the history of his invention, but not much about the difficulties he experienced in getting recognition for it. To the question of why he was silent about being ignored for so long, he explained to me that he had made protests and filed complaints, but with no results.

Furka is not the type of person to fight for his rights. Although he thrives in any endeavor where his creative mind has a role, he freezes easily in worldly pursuits. His friends used to prod him to write a book about combinatorial chemistry, and he finally did.

Then it was suggested that he find a publisher for it. Furka wrote to a total of one (!) publisher, which did not bother to respond. Furka gave up writing to publishers and instead made his manuscript available on the Internet.[9]

Furka's experience shows that great ideas may be born even at the peripheries of science, but it also shows that getting proper recognition is much harder when such ideas appear outside the centers of scientific research. Furka likes to point out that the principles he launched are applicable to fields in chemistry other than just peptide synthesis, as well as in other branches of sciences and technologies. He is still full of ideas, but he has other interests as well. He likes the outdoors, he proudly produces his own wine in his vineyard, and he enjoys his travels to various corners of the world. The split-mix method has reached chemistry laboratories all over the world; certainly in more places than Furka's tourism has taken him.

Figure 7.1. Rosalyn S. Yalow in the laboratory in 1948.
Courtesy of Rosalyn S. Yalow and Benjamin Yalow,
New York City.

CHAPTER 7
PROVING HERSELF
Measuring Hormones in Blood

The world cannot afford the loss of the talents of half of its people if we are to solve the many problems which beset us.
Rosalyn Yalow[1]

Rosalyn S. Yalow (1921–2011) suffered from multiple handicaps at the start of her career, but she did not let any of them stop her. She came from a poor family, she was Jewish (at that time, it was still difficult for Jews to get into academia), and she was a woman. But times were changing in the 1940s.

Yalow became a physics PhD and got a job at the Bronx Veterans Administration Hospital. There she developed a fruitful collaboration with Solomon A. Berson (1918–1972), an outstanding MD, and the two jointly made a plethora of discoveries in biomedicine. Their best-known discovery is a new technique called radioimmunoassay, which allows for the determination of extremely small amounts of hormones and other vital substances in the blood that used to be impossible to measure. Yalow and Berson's discoveries were not well received at first, but gradually they came to be accepted. When Berson died, Yalow had to continue alone. She persevered and was awarded the Nobel Prize in Physiology or Medicine. Although Yalow paid a price in human relations for her success, she was satisfied that she had proved herself to the world—and to herself.

*   *   *

Rosalyn Yalow died on May 30, 2011, just as this chapter was being readied for publication.

During the eighteen years of their collaboration, Solomon Berson and Rosalyn Yalow were inseparable in the laboratory, and their achievements were indistinguishable. They had great respect for each other and each had knowledge that complemented that of the other. They had different personalities, but when Berson died in 1972, Yalow forced herself to take over some of Berson's functions in her efforts to bring to completion everything they had achieved together.

I visited Rosalyn Yalow in the spring of 1998. When I had been corresponding with her about my visit, I had no idea that she had been partially paralyzed by a series of strokes.[2] During our conversation, however, I soon forgot about her condition: her willpower appeared stronger than her incapacitation, and her fierce determination shone through her story.[3]

When I left Yalow's office at the Veterans Affairs Medical Center, I sat down in the hallway to take additional notes about my impressions. I was still there when a helpless older woman was wheeled out of her office. I hardly recognized Yalow; she was so different from the person I had just finished speaking with. At the time, Yalow was considered a medical miracle because she had survived what would have killed people of lesser willpower. A dozen years later, she still keeps going.

Yalow came from a poor Jewish immigrant family and had to overcome various obstacles to receive the education she craved in order to find a job in which she could do scientific research. She and Berson had built up an exemplary partnership in science, but his premature death might have left her to disappear in oblivion. Berson was considered the principal person of their team: he was older, he was a physician—their discoveries were biomedical—and he was male. There was no doubt that, had Yalow been the one to die prematurely, Berson would have faced no obstacles in achieving increasing recognition.

Berson was born in 1918 in New York City to an immigrant family from Russia. He studied at City College of New York and graduated in 1938. He wanted to become a medical doctor, and he applied to twenty-one medical schools, but none accepted him. This was not at all unusual during the time of anti-Jewish discrimination in American medical schools. Berson ended up studying science at New York University, where he was then admitted to its medical school in 1941. He became an MD in 1945. Following an internship and army service, he joined the Bronx Veterans Administration Hospital for training in internal medicine as part of his plan to become a practicing physician.

In the spring of 1950, Bernard Straus, chief of medicine at the hospital, suggested that Berson talk with Rosalyn Yalow, a nuclear physicist who was expanding the radioisotope service at the hospital and who was looking for a physician to work with. Yalow had never even taken a course in biology, but she wanted to proceed with medical research. It was a bizarre situation in that Berson was practically interviewing with a person who was looking for her potential boss, and the interview became a mutual test of each other's abilities. But something clicked, and the two became lifelong partners in research. They fortunately complemented each other's backgrounds: she had physics and he had medicine and biology. They were equally driven, but they did not compete with each other. Fortunately, each had a supporting family. Of the two, he was the more visible and the more vocal, and in 1954 he was put in charge when their unit became the independent Radioisotope Service of the hospital. The situation suited Yalow because her interest was strictly in research; besides, Berson was a physician, and the unit had clinical responsibilities as well.

Yalow was born as Rosalyn Sussman. Her maternal great-grandparents came from a well-to-do family in Germany. Their daughter, Rosalyn's maternal grandmother, Bertha, was considered the black sheep and had to leave the family, so she immigrated first to Riga (then in Russia, now Latvia), and eventually to the United States. Her offspring turned out to be the lucky ones; the descendants of the rest of the thriving family perished in the Holocaust. The American wing of the family, including Bertha's daughter, Clara, who became Rosalyn's mother, remained uneducated and poor. Rosalyn's paternal family was also poor; they had come from Russia, although her father was born in New York. Rosalyn's parents worked hard and succeeded to some extent, only to have their lives adversely effected by the Great Depression. Still, they wanted to give their children an education, although they did not set their sights higher than for Rosalyn to become a schoolteacher.

Yalow, however, aimed higher. She was very determined from early childhood and developed an interest in mathematics and the sciences. She made good use of the public library and was lucky to have an inspiring chemistry teacher in high school. She went to Hunter College—at that time a women's college, the equivalent of which for men was the City College in New York City. Both schools were tuition-free, but the academic requirements for acceptance were tough. Perhaps the largest number of future Nobel laureates in the world has graduated from this college system. Yalow was inspired by three physics professors and by the science of nuclear physics, in which new discoveries were being made literally in front of her eyes. Hunter College did not have a physics department at that time, but Yalow pushed herself to learn. She attended a lecture about nuclear fission by Enrico Fermi at Columbia University in January 1939, just a few weeks after its discovery!

Following graduation, Yalow took a part-time secretarial job with the biochemist Rudolf Schoenheimer at Columbia University, as it was unlikely that she would be offered a scholarship for graduate school in spite of her excellent academic record. Schoenheimer was a pioneer in applying radioactive isotopes in the study of biological processes; he labeled amino acids and followed their movements in the organism during metabolism. He wrote a very influential book, *The Dynamic State of Body Constituents*;[4] alas, by the time the book was published in 1942, Schoenheimer had committed suicide (in 1941). Yalow had left after a few months, in the summer of 1941, with no real scientific interaction between them. She had been offered a teaching assistantship and a chance at graduate studies in physics at the University of Illinois at Urbana–Champaign.

Yalow arrived at the university in September 1941, where pivotal changes soon took place in her life. She met her future husband, Aaron Yalow, and she was taken on by a top physics professor, Maurice Goldhaber, as was Aaron. Goldhaber was a refugee from Germany. He authored research papers, which had to be classified for the duration of the war for their strategic importance. The Manhattan Project used Goldhaber's results, but he was not invited to join the organization because he lacked the necessary clearance, as his in-laws lived in Germany. By the time the Manhattan Project was under way, Goldhaber's in-laws had been murdered by the Nazis, but Goldhaber and his wife learned of this only after the war.[5]

Yalow was the only woman among four hundred faculty members and graduate students at the engineering faculty at Illinois, but she stood her ground. Goldhaber characterized her to the other professors as rather aggressive, but Goldhaber noticed that his American colleagues construed it as compliment even though he did not mean it as one. Goldhaber had had firsthand experience with determined women scientists, as his wife was one, too.

Rosalyn was lucky that she found Aaron; he was ideally suited for her. She became a devoted wife and later a devoted mother, but she made it clear from the start of their married life that she cared for her career to the extent few wives at that time did, and that Aaron's career had to be subordinated to hers. This was fine with him; he did not plan to go into research; rather, he was happy to be a physics professor and found pleasure in advising his wife. It was he, for example, who suggested that she enter nuclear medicine, an emerging field at the time, so she could expect to compete there successfully.

Aaron and Rosalyn got married in 1943. For Aaron, it was an essential part of family life to observe the Jewish religion in the orthodox way. Rosalyn did not come from a highly observant family and recognized only the traditional holidays, but now she set herself to keeping kosher and leading an observant household. This was her contribution to another of marriage's compromises. But here, again, Aaron was magnanimous. Whereas at home everything was strictly followed, it was also understood that she would be free to work in the lab on Saturdays and that she would need to drive there and back, an act that would be abhorrent to many orthodox worshippers.

Rosalyn was an excellent student and finished her graduate studies before Aaron did. She immediately moved back to New York in January 1945, where she went to work as a physicist at a research laboratory for International Telephone and Telegraph Corporation (IT&T). Soon Aaron joined her in New York and eventually landed a job at Cooper Union College. When the IT&T laboratory moved from New York, Rosalyn quit and started teaching physics at her own alma mater, Hunter College, which now had not only women students but also returning male veterans. Teaching was not her main interest, but she did it as conscientiously as she did everything else. She found much joy in a few dedicated students who wanted to learn physics, among whom was the now famous Mildred Dresselhaus.

Yalow helped Dresselhaus to become a scientist, and their interactions have lasted throughout the decades. Dresselhaus knew a different Yalow behind the rigorous and cold facade, finding that she could be "very motherly."[6] When Yalow knew that Dresselhaus was busy preparing a talk, "she would bring a shopping bag full of stuff, a little like a housewife."[7] For Dresselhaus, Yalow was a very good role model in spite of the differences in their personalities: "I'm a lot less aggressive than she and much more accepting. I've put a lot of time into issues of women and science and she doesn't really like that, but she thinks that I've done the right thing and she supports me. She wouldn't do it herself . . ."[8] Yalow never believed in the women's movement, though she always believed in women's ability to be equal with men in science as elsewhere.

Aaron helped Rosalyn to meet Edith Quimby, a noted radiation scientist at Columbia University College of Physicians and Surgeons. She and her boss, Gioacchino Failla, founded the Radiation Research Laboratory at Columbia. First, Yalow volunteered to work in Quimby's laboratory. Then, Failla helped Yalow to become a part-time consultant at the Bronx Veterans Administration Hospital. This was in December 1947; in 1950, Yalow's position turned into a full-time one. Yalow set up her laboratory in a janitor's closet with meager support, but she did a lot of collaborative work and published papers during her time there.

The year 1950 was a fateful one. Yalow and Berson met and began an eighteen-year collaborative relationship until Berson left in 1968. Their interactions continued for four more years until Berson's untimely death in 1972. The Yalows had a live-in maid, and Rosalyn's mother came to help every day when the children, first Benjamin, born in 1952, then Elanna, born in 1954, were small. Otherwise, Yalow's life revolved—literally day and night—around her research. In those days, women were forced to resign from their jobs at the fifth month of their pregnancy (not just take leave, but resign!). Yalow did not want to do that, and because Berson was in charge, nobody could force her.

The Veterans Administration was supportive of basic research, which Berson and Yalow initially did, in spite of the fact that it did not seem to be directly related to their duties. Eventually, though, that research brought tremendous benefits in biomedical applications. Still, their laboratory was not exactly in the mainstream of research at the time, which actually had its merits since its backwater status shielded Berson and Yalow from undue competition and intrigues. They could work on problems of their own choosing and develop their work at their own pace. Not that this pace was slow; on the contrary, they were driving themselves with no restraint, preferring to carry out mass experiments themselves even when those experiments could be entrusted to others.

Berson and Yalow's research projects were diverse, but the

Figure 7.2. Solomon Berson and Rosalyn S. Yalow in 1957.
Courtesy of Rosalyn S. Yalow and Benjamin Yalow, New York City.

application of isotopes was common among them. (Isotopes are variants of the same element, differing in the numbers of neutrons in their nuclei; they are often radioactive.) It was as recent as 1944 that George de (or Georg von) Hevesy, the Hungarian-German-Swedish chemist, was awarded the Nobel Prize (for the year 1943) for work that bore direct relevance to Berson and Yalow's research. The motivation for Hevesy's award read: "for his work on the use of isotopes as tracers in the study of chemical processes." The roots of Hevesy's discovery have been described in various anecdotes, so let us retell just one here, because it provides a graphic way to understand the essence of the technique.

Hevesy was staying in a German boardinghouse and noticed that when there was beef for dinner one evening, and pieces of it were left on the plates, the soup next day contained similar meat. He suspected that their hostess might have recycled the leftover meat, and he was looking for a way to confirm this. At the time, he was working with radioactive isotopes, so one evening he left a piece of meat on his plate and smeared on it a little harmless substance containing radioactive isotope. The next day Hevesy tested the soup with his Geiger–Müller counter, and it showed radioactivity.

Hevesy pioneered the use of radioisotopes to measure blood volume in 1940. The determination of the volume of blood in an organism is a simple one that uses the principle of tracer dilution. A detectable label (here, a radioactive material, though it could also be a dye) is introduced into the bloodstream in a well-defined amount. After sufficient time to achieve complete mixing in the organism's bloodstream, a sample of blood is taken from the organism, and its radioactivity is measured. Let's say an amount of radioactivity of 1,000 units had been introduced, and now a sample of 10 milliliters (one-third of a fluid ounce) is taken, and its radioactivity is measured to be 2 units. Thus the amount of radioactivity in 10 milliliters is 500 times less than the original amount.

That means that the total volume is 500 times larger than the volume of the sample; that is, 5,000 milliliters, or 5 liters (five and a half quarts), is the total volume of the blood in that organism.

Berson and Yalow carried out an extensive program in which they used radioactive tracers for a variety of physiological problems. One of their favorite tracers initially was radioactive iodine, ^{131}I. This notation contains the chemical symbol of iodine, I, and in front of it (or sometimes behind it), the atomic mass of this iodine isotope, 131, is given as a superscript. The atomic number of iodine is 53, meaning that iodine—any isotope of iodine—has 53 protons in its atomic nucleus. The number 53 need not be indicated, because by giving the symbol of iodine, it is unambiguous that the atomic number is 53. In contrast, the atomic mass must be given if we want to know which isotope of iodine is being referred to. The natural iodine is ^{127}I; that is, its atomic mass is 127, but it would be useless in tracer studies, because it is not radioactive. The isotope ^{131}I does not occur in nature; it is produced in nuclear reactors. The iodine isotope ^{131}I has a rather high energy of emission associated with its radioactive decay. As are all isotopes of iodine, it is taken up and stored in the thyroid gland. Because of its high energy, ^{131}I is used medically to depress thyroid tissue. This usage was not important to Berson and Yalow, hence another iodine isotope, ^{125}I, which emitted much less energy and was therefore much safer to work with sufficed for their measurements. So Berson and Yalow very quickly turned to ^{125}I as their tracer isotope of choice.

In addition to the determination of blood volume, one of the first applications of radioisotopes in Berson and Yalow's work was the clinical diagnosis of thyroid diseases and the mapping of what happens to iodine in the human organism. Berson and Yalow's work in the area of thyroid function was groundbreaking and put them on the map in terms of being recognized as bright young investigators. But their first big discovery was related to their studies of diabetes.

Aaron suffered from diabetes, so Yalow had some familiarity with the disease, but there is no indication that she engaged in diabetes research because of her husband's illness. Berson was a fully trained endocrinologist. Endocrinologists deal with, among other things, the hormones—such as insulin—needed to enable digestion of carbohydrates and fats. The sugar levels of diabetic patients are too high; this is caused by insufficient amount of insulin (made by the pancreas) that is supposed to break the sugar down in order to produce energy. The initial project involved checking a hypothesis of I. Arthur Mirsky (no relation to Alfred Mirsky, mentioned in chapter 4), who did a lot of research on the metabolism in diabetes. According to Mirsky's hypothesis, the normally secreted insulin degraded far too rapidly in diabetes patients who had developed the illness as adults. Yalow and Berson administered labeled insulin from the pancreas of pigs or cattle to three groups of people, including diabetes patients who had already been under insulin treatment for some time, patients who had never had diabetes but who had received insulin to produce hypoglycemic shock as a treatment for schizophrenia, and healthy individuals without any history of insulin treatment.

The observation was that insulin was retained longer in the bloodstream of patients who had been treated with insulin than in the bloodstream of those who had not. This was true regardless of whether the patients were diabetics or were suffering from schizophrenia rather than from diabetes. Most importantly, the team found that in the insulin-treated patients radio-labeled insulin was bound to a protein, globulin.

Berson and Yalow's conclusion was that the body behaved toward insulin as it does toward any foreign antigen and developed an antibody to bind it. This was independent of whether the patient was diabetic or whether he or she suffered from something else. The common denominator was the prior insulin treatment. The negative consequence for diabetics was having insulin in their blood that was bound to a much larger molecule and therefore was not available for performing its physiological function. This was a major discovery: that insulin induced the formation of an antibody.

Berson and Yalow overturned a long-held dogma according to which insulin (at about 6,000 Daltons—atomic mass units) was too small to induce the production of antibodies. Their discovery then had to overcome all the hurdles encountered by any new discovery that overturns long-held dogmas. Fortunately, Berson and Yalow's paradigm change was accepted fairly quickly by the scientific community, but this did not happen without a fight. Their initial manuscript was rejected by *Science* and also by the *Journal of Clinical Investigation*. The team had to revise the manuscript, and encountered a final dispute over the title of their communication. According to the September 29, 1955, letter from the editor in chief of the journal, "The experts in this field have been particularly emphatic in rejecting your positive statement that the 'conclusion that the globulin responsible for insulin binding is an acquired antibody appears to be inescapable.' They believe that you have not demonstrated an antigen-antibody reaction on the basis of adequate criteria, nor that you have definitely proved that a globulin is responsible for insulin binding, nor that insulin is an antigen."

The original title of the manuscript was "Insulin-I[131] Metabolism in Human Subjects: Demonstration of *Insulin Transporting Antibody* in the Circulation of Insulin Treated Subjects." Here I italicize the phrase that the journal found unacceptable. The paper was finally published with this title as a compromise: "Insulin-I[131] Metabolism in Human Subjects: Demonstration of Insulin Binding Globulin in the Circulation of Insulin-Treated Subjects."[9] Thus the phrase "Insulin Transporting Antibody" was replaced with "Insulin Binding Globulin."

The controversy over this publication left a scar on Yalow. She was a person who held grudges, which she vented when it would hurt the most. It is not at all unprecedented that journals reject paradigm-changing discoveries, but it is seldom that fresh Nobel laureates bring up their feelings about such rejections in their Nobel lectures. Yalow did just that, repeating the damaging excerpt of the letter of rejection in her Nobel lecture with a vengeance.[10] The editor in chief, Stanley E. Bradley, was a reputable scientist on the faculty of Columbia University who spent his entire life in education and research. He was, of course, relying on the reviews submitted to him by "experts," who must have been immunologists and whose names have remained anonymous, as the names of reviewers should.

The main reason Berson and Yalow's discovery gained acceptance was the practical utilization of their works that appeared in a most useful methodology called *radioimmunoassay*, or RIA. The observation about the labeled insulin was that it was bound to the antibody and that part of it was displaced from its bound position. It was then possible to measure the amount of the displaced radioactive insulin in a well-defined volume of the blood of the patient. This amount was compared with the displacement amounts produced by known concentrations of insulin preparations administered to the patient. This way, the team calibrated the displaced amount of labeled insulin and so had a technique to measure the concentration of insulin in the blood. The name of this new technique incorporated the words *radio* (the application of radioactive isotopes), *immuno* (the use of immunological response), and *assay* (the fact that the procedure was a measurement).

The principle was simple, and the application was universal. The antigen/antibody interaction was specific, so any hormone, or other small molecule for which an antibody response could be specified, became suitable for applying RIA to its quantitative determination. The technique needed to be very sensitive, and it was, because hormones, like insulin, are present in the organism in extremely small concentrations. How small? Imagine a typical insulin solution in which there might be just one insulin molecule among 55,000,000,000,000—fifty-five thousand billion—water molecules. Finding a needle in a haystack would be child's play compared with finding the insulin molecule in water at such a low concentration. But this is exactly the kind of concentration to which RIA has been applied to perfection.

Nonetheless, the medical community was slow to accept RIA, perhaps because it seemed improbable, if not impossible, to carry out such a sensitive measurement. Berson and Yalow did everything they could to popularize their technique and to educate potential users on how to apply it. They invited hundreds of medical scientists to their laboratory and showed them how to perform RIA, and even gave them sufficient supplies of antibodies so they could try them out in their own laboratories. Berson and Yalow's interest was purely in the advancement of science. Others suggested that the team patent RIA, but Berson and Yalow emphatically rejected that idea. They were interested only in having the technique used widely, and eventually this came to fruition.

As long as Berson and Yalow worked together, Yalow cared very little about the outside world because that was Berson's domain. When Berson left the Veterans Administration in 1968, and especially when he died in 1972, Yalow was left alone, and all the work they did together fell onto her shoulders. In addition, everything that had been Berson's domain of activity, such as most of the writing and almost all the lecturing, became Yalow's responsibility. She was also aware of the fact that her peers and many others were watching her closely, and not always sympathetically. They watched to see whether she would be able to keep up the scientific production from her laboratory or whether there would be a sharp decline in production. A decline in production would have been understandable, but Yalow knew she could not afford it.

On top of all the extra work and scrutiny, there was the expectation of the Nobel Prize. Yalow and Berson had been nominated for the prize shortly before Berson's death, but the stipulations in Alfred Nobel's will do not allow awarding the prize posthumously. Yalow's situation had now become very difficult. Berson had been the official boss of their laboratory, and most people had assumed that Berson was the brain behind their joint discoveries. He was more visible; he was witty and popular; and he was a man.

The scientific world used to be the man's world, and even though things have changed, this is by and large still so. When Berson and Yalow went to meetings in the 1950s and 1960s, it was not very common for women scientists to be in attendance. Those women who did attend were mostly the wives of scientists for whom separate "ladies' programs" were organized. Even Berson found it difficult to deal with Yalow's presence at these meetings, and sometimes he advised Yalow to sit with the wives rather than with the scientists at the social events of these conferences. It was a truly awkward situation for Yalow, but as long as her contributions were properly represented, she did not mind. Now, all this was to change.

By all indications, Yalow grew up to the challenge she was facing. It was a beautiful idea to dedicate their laboratory to Berson's memory by naming it after him. Yalow stated that she had requested this designation "so that his name will continue to be on my papers as long as I publish . . ."[11] Yalow must have felt that she had to find a delicate balance between assigning credit to Berson and maintaining her own recognition. It would appear that she resolved this issue impeccably. The most convincing argument for her own scientific acumen is the extraordinary research results that she produced alone or in collaborative efforts with her associates. The period between Berson's death in 1972 and the Nobel Prize announcement in 1977 was a most trying time for Yalow.

Yalow's Nobel Prize was "for the development of radioimmunoassay of peptide hormones." Rarely has anyone been as conspicuously missing as Berson was for this award. At the award ceremony, Swedish academician Rolf Luft described the achievements of the new laureates in physiology or medicine, of which there were three. Yalow received half of the prize, and Roger Guillemin and Andrew V. Schally shared the other half for their work on the production of peptide hormones in the brain. Luft was very eloquent in his presentation speech; referring to the Berson–Yalow technique of RIA and its tremendous impact on research and medicine and on many lives. He found it appropriate to paraphrase Winston Churchill's comments about the Royal Air Force pilots during the Battle of Britain in 1940. Luft declared, "Rarely have so many had so few to thank for so much."[12]

Luft truly appreciated Yalow's contribution and perhaps her personality as well. He expressed his admiration by giving her a smaller gold replica of the Nobel medal (he also gave an identical medal to his wife). Yalow was not given to wearing jewelry, but she liked the medal. Some have criticized her for wearing this Nobel memorabilia; some have even thought she'd had it made for herself.

The replica story is but a minor example of how Yalow's every action, whether real or perceived, invited scrutiny and even contempt. This type of reaction was compounded by her increasingly tactless behavior during the years following the awarding of the Nobel Prize. Perhaps the many years of terrible tension gave way, but if she had not been particularly popular before, her estrangement from much of the community in which she functioned as a scientist—and this was her only community—intensified. Then came the debilitating strokes and other accidents, but her extraordinary willpower and determination to prove herself and never give up—qualities that had been driving her all her life—have remained intact.

CHAPTER 8
STUBBORNNESS
"Impossible" Matter

. . . people can find me tough.
Dan Shechtman[1]

Dan Shechtman (1941–) made a serendipitous observation, which became a triumphant discovery—thanks to his stubbornness and perseverance. These were important traits. When Shechtman was a student, he passed an exam in crystallography by giving proof of why fivefold symmetry was impossible in the world of crystals, which was a long-held dogma in condensed phase science. Shechtman's discovery of quasicrystals helped to destroy this dogma. Alas, the most influential chemist of his time declared Shechtman's discovery a nondiscovery. The journal to which Shechtman submitted his first report about his discovery rejected it, saying that physicists would not be interested in it. When Shechtman finally succeeded in bringing out his report, an avalanche of publications followed. Shechtman's discovery has proven important for chemistry, physics, materials science, and even for design science and the arts.

I t used to be a generally accepted belief that fivefold symmetry was impossible in the world of crystals, where atoms or molecules build up a structure and there is neither overlap nor any gap in their packing. Here is a simple illustration of what is meant by

Figure 8.1. Flowerlike icosahedral quasicrystals in quenched aluminum-manganese sample (by Ágnes Csanády, Budapest, and Hans-Ude Nissen, Zurich). Courtesy of Ágnes Csanády.

the impossibility of fivefold symmetry in crystals: We can use same-size regular triangles, squares, and hexagons to cover a surface without gaps or overlaps, but it is impossible to do so with same-size regular pentagons. Similarly, sevenfold and higher symmetries used to be considered impossible for this purpose as well.

For a long time, illustrious scientists and artists such as Johannes Kepler and Albrecht Dürer tried to create patterns in

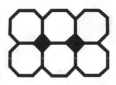

Figure 8.2. Same-size regular triangles, squares, etc. Only the triangles, squares, and hexagons are capable of covering the surface without gaps or overlaps.

which regular pentagons covered a surface, but they did not succeed. Then, the British mathematician Roger Penrose gave it a try.[2] Penrose came from a family of unusual interests. His father, Lionel Penrose, held the Galton Chair of Human Genetics at Oxford University, and Lionel instilled his interest in science in his son. Roger Penrose has since become not only a noted mathematician but also a celebrated author. One of his pastimes, especially when he was at meetings that bored him, was doodling.

One day, a logo on someone's letterhead caught his attention. The logo was composed of a pentagon in the middle, surrounded by five other pentagons within a larger pentagon. This pattern induced Penrose to extend it; that is, to continue the pattern by creating larger and larger pentagons. He used not only the whole pentagons but parts of them, too, to cover the surface of the paper. It led to an appealing pattern: the page was covered with regular pentagons, but the shapes were of gradually changing sizes. This came as close

Figure 8.3. (left): Roger Penrose in 2000 at Oxford University, United Kingdom. Photo by the author. Figure 8.4. (below): Pattern of regular pentagons with changing sizes.

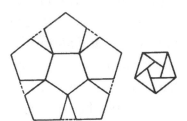

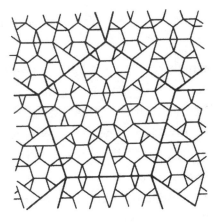

to Kepler's and Dürer's dreams as Penrose could have hoped for. He communicated his finding in an obscure mathematical journal, but when mathematics and science writer Martin Gardner wrote about it in *Scientific American*,[3] Penrose's patterns became widely known.

Another British scientist, crystallographer Alan Mackay, was intrigued by Penrose's pattern. Crystal structures can be investigated by employing the diffraction phenomenon in which X-rays, neutrons, or electrons shined onto the substance change their directions, and an interference pattern results from which the structure may be determined. Mackay decided to try to simulate a diffraction experiment from the Penrose pattern. He figured that if he succeeded in producing a diffraction pattern displaying sharp bright spots, that would be an indication that even three-dimensional structures could exist that would correspond to what Penrose did on a sheet of paper.

Mackay was very alert to structures that fell beyond the seemingly perfect crystals. He followed in the footsteps of his mentor— J. Desmond Bernal—in his attempts to broaden the scope of crystallography. Mackay succeeded in producing his simulated diffraction experiment and issued a warning to other scientists in his lectures and articles that there might be structures of fivefold symmetry, but we might miss noticing them because the dogma of their nonexistence has been ingrained strongly in our science.[4]

I had been intrigued by Mackay's publications and was happy to meet him in 1981 in Ottawa, at a crystallography meeting. We started corresponding, and he visited us in Budapest in September 1982. He gave three lectures, two of which he devoted to various aspects of fivefold symmetry. It was on this occasion that I heard his warning about the possibility of solid structures with fivefold symmetry. His views sounded esoteric, but he made them seem attractive. Nobody in the audience—not even Mackay—knew that, by then, structures that Mackay predicted might exist had been observed experimentally five months before.

Unbeknownst to Mackay and Penrose, Dan Shechtman was doing experiments that had a direct bearing on what Penrose and Mackay were trying to do. Shechtman, a graduate of the Technion (Israel Institute of Technology) was a materials scientist who studied various substances and prepared new ones with desired properties, among them new metallic alloys. He worked at the

Figure 8.5. Dan Shechtman and Alan L. Mackay in 1995 in the Hargittais' home in Budapest. Photo by the author.

Technion but often spent stints of various lengths at American research laboratories.[5] He was not at all concerned with the problem of fivefold symmetry.

In 1981, Shechtman took a sabbatical from the Technion at the

US National Bureau of Standards (NBS; now the National Institute of Standards and Technology), near Washington, DC. He was invited by one of its senior scientists, John Cahn, to spend some time at NBS, because he had developed a new technique for studying metallic powders in the electron microscope. Shechtman's research at NBS was sponsored by the Defense Advanced Research Project Agency (DARPA), but he was free to strike out in any direction he felt would be interesting.

Since Shechtman wanted to create new alloys, he started studying rapidly solidified aluminum–iron alloys and examined how the composition and the conditions of their solidification impacted the structure and properties of the new alloys. Rapid solidification is one of the approaches used to influence the properties of alloys. He learned a lot about the process of rapid solidification and prepared several publications with his colleagues at NBS. They made useful materials, but the technique they developed did not become a widespread technology. The discovery Shechtman made was unexpected and was a by-product of these studies.

In some of his experiments, Shechtman wanted to compare two alloys of similar composition: one consisted of aluminum and iron, and the other of aluminum and manganese. As part of these experiments, he prepared a series of aluminum–manganese alloys with increasing amounts of manganese. For practical purposes he had to limit the manganese content to a few percent; otherwise the alloy would become brittle and, therefore, useless. At one point, however, Shechtman's curiosity overtook his practical sense, and he kept increasing the amount of manganese in the alloys. This is the kind of experimentation we read about in publications only if it leads to results; experiments that do not, disappear.

On April 8, 1982, Shechtman was studying in his electron microscope a sample of rapidly solidified aluminum–manganese alloy that contained 25 percent manganese, when he observed something utterly unexpected. The diffraction pattern he recorded on a photo-graphic plate showed distinct features that could only be interpreted as originating from a structure having tenfold symmetry. He duly recorded in his journal for the plate number "1725, Al–25%Mn" the following: "10 Fold ???" He found ten bright spots in the diffraction pattern he was examining, and they were equally spaced from the center and from each other. He counted them, counted them again, and then said to himself in Hebrew, "*Ein chaya kazo*" (There is no such animal). He was sufficiently versed in general crystallography to recognize at once that he had found something extraordinary. He did not know about Mackay's simulated experiment, let alone Mackay's warning. Although he was slightly familiar with Penrose's pattern, Shechtman did not make the connection between his work and the pattern until long after his discovery.

At the time of his crucial experiment, Shechtman was alone in the laboratory, but he felt the urge to share his excitement with somebody else—this is a common feeling among discoverers at such a moment. Shechtman walked out into the corridor, but nobody was there, so he returned to his electron microscope and performed a series of additional experiments. On this single occasion, he essentially performed all the experimentation he would need in order to be certain of his observations. In a few days, every-thing would be ready for the announcement of the discovery. But it would take more than two years to publish it.

For quite a while, Shechtman was alone in his conviction that he had actually discovered something. After conducting his exper-iment, he started inquiring at NBS what his colleagues knew about tenfold symmetry, but instead of meaningful responses, he was met with ridicule. The kind ones tried to explain to him that he must have seen some other phenomenon, and someone even gave him a textbook on X-ray crystallography to help him understand why what he thought he had seen was impossible. Various explanations

Figure 8.6. (left): Dan Shechtman in 2007 in Budapest. Photo by the author. Figure 8.7. (below): A page in Dan Shechtman's lab journal, April 8, 1982. Courtesy of Dan Shechtman, Haifa. Note the remark at the entry for the pattern no. 1725: 10 Fold ???

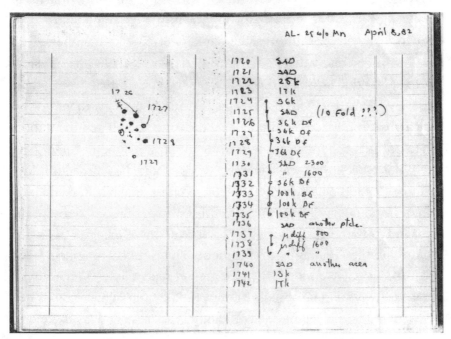

for the origin of his pattern were offered, including the phenomenon of crystal twinning, which is when two crystal domains grow
together and produce unusual symmetries. An alternative thought
was that there might be crystal defects that would produce the pattern of impossible symmetry. The irony of the situation was that
when Shechtman was a student, he had to show the proof that fivefold symmetry was impossible in crystals.

Dan Shechtman was born in Tel Aviv. His mother's family had
arrived during the second wave of immigration, in 1906, and his
father's family with the fifth, in 1930. Both sides originally came
from Russia. Shechtman went to primary school in Ramat Gan, a
suburb of Tel Aviv, and when he was fourteen years old, his family
moved farther away, to Petach Tikva. Because their homes were
always crowded, Shechtman would later have his own house built
with plenty of space and separate rooms for each of his four children.

In his youth, Shechtman was involved in the Zionist-Socialist
movement, but he never even flirted with communist ideology. He
appreciated that the movement taught him important values and
shaped his character through the camaraderie of its members. The
group enjoyed many different kinds of activities, including physical
exercises and weeklong field trips in the desert that tested their
endurance under severe conditions. Shechtman had a mentor who
was a great influence on him, but by the time he realized the importance of his teacher and wanted to thank him, his mentor had died.

When Shechtman was eighteen years of age, he entered military
training. After the basics, he was selected for a special course in psy-
chology and interviewing. He used this training during the rest of his
two and a half years of military service. In 1962, he began his
studies at the Technion to become a mechanical engineer—his
dream from early childhood when he had read a Jules Verne novel.
He was always a good student but never at the top of his class. When
he graduated with his bachelor's degree in 1966, he was invited to

continue his studies for a master's degree. He earned his keep by instructing students and at the same time began doing research.

Shechtman took all the necessary subjects toward his degree: metallography, X-ray crystallography, and electron microscopy, among others. When the first electron microscope arrived at the Technion, Shechtman was among the very first students who became an expert in working with it. He stayed on for his doctorate after he had completed his master's studies. Titanium alloys and their properties—an important topic for aviation—became his doctoral project. When he received his PhD degree, he wanted to stay at the Technion, but he was told that first he should have some years of experience abroad.

In 1972, Shechtman sent letters of inquiry to a hundred universities and research institutions around the world. Of the two offers he received, he accepted a position at the US Air Force Research Laboratories at Wright-Patterson Air Force Base near Dayton, Ohio. He studied titanium–aluminum alloys for the next two and a half years. He and his family considered staying in the United States because there were no offers from Israel, but, finally, the Technion decided to employ him. Shechtman and his family returned to Israel in 1975, and he took a position as a lecturer in the section of metallurgy within the Department of Mechanical Engineering, which eventually became an independent department of materials engineering.

Shechtman was not afraid to give away any secrets from his work; on the contrary, he tried to talk with anyone who was willing to listen to him about his experiment. He put his "impossible" diffraction pattern on a Christmas card to help disseminate the information. (His DARPA sponsor had this Christmas card on the wall of his office.) Shechtman did not know what else to do; he had no further experiments in mind that would make his case stronger, because, in his mind, what he had shown was unambiguous. He put the alloy sample under his electron microscope from time to time to look at the pattern again, as if checking to see if it was still there. It was.

When his stint at NBS was over, Shechtman returned to the Technion and continued looking for colleagues who would be willing to engage in a meaningful discussion with him. Finally, he found one, Ilan Blech, who was an expert in X-ray diffraction. The two started making models of structures that might produce the diffraction pattern that Shechtman had observed. They built a model that consisted of icosahedral parts; the icosahedron has fivefold symmetry. As they produced it, their line of thoughts started converging toward that of Mackay's, who, two decades before, had published an icosahedral structure that was not part of classical crystallography.[6]

Blech's support boosted Shechtman's spirits. There was at least one scientist who shared his conviction that, however unexpected the results his experiments yielded, they were real. Because nobody else believed in what Shechtman did, he had not attempted to write up his observation in a research paper. He was in a precarious situation. At NBS, his group leader felt that he had to protect the reputation of his group by resisting Shechtman's claims and so expelled Shechtman from his group. At the Technion, Shechtman was not yet a professor and did not feel that the atmosphere was sufficiently supportive to take a step that would have been considered risky by most.[7]

A more established scientist might have rushed to publish such a new finding lest it be scooped by others. On the other hand, a more established scientist might have been very cautious, too, in coming out with a discovery that others thought impossible. Shechtman was convinced that he had a novel observation, but it bothered him that he could not offer a credible explanation for it, and so he felt reluctant to try to publish it. After a while, he and Blech offered a model of a structure that produced a diffraction pattern in the computer like the one Shechtman had observed.

When Shechtman finally decided to submit a manuscript in which he and Blech described his experiment, it was done in a tentative manner. The information was there, but an unsuspecting reader might not have noticed it, since it was buried under a mountain of information about the aluminum–manganese system. The manuscript read like a report on metallurgy, yet the team had submitted it to a physics publication, the *Journal of Applied Physics*—and the editor of the journal did not seem to be sensitive to whatever subtleties existed in the manuscript. He wrote back almost immediately that the material in the manuscript was not suitable for the journal and—incredibly—that it would not interest physicists. It might be interesting to know what this editor thought when, after a couple of years there was an avalanche of papers by physicists, among others, once Shechtman finally managed to publish his experiment.

Shechtman and Blech's manuscript was not composed in a straightforward manner, so they are also somewhat responsible for the failure of the publication of their manuscript. However, it is not that rare that such seminal discoveries are difficult to publish. Bad papers are expected to be rejected, but often this is the fate of manuscripts that report radically new results. This conservatism appears in the workings of the most prestigious journals in a most conspicuous way. As the Shechtman-Blech manuscript was indeed more a report about studies in metallurgy than in physics, the team decided to resubmit it to a metallurgical periodical. It was eventually published, but it had practically no impact on future developments.

Shechtman knew that he not only needed a plausible explanation for the phenomenon he observed but also that he needed to considerably improve his presentation. This is why he turned for help to John Cahn, a veteran scientist in the section where he worked at NBS. Cahn had known about Shechtman's discovery, but for two long years he did not believe in its veracity. Now he changed his mind, and the grateful Shechtman invited Cahn to be a coauthor of the report. Then Cahn involved yet another participant, Denis Gratias, a young French mathematical crystallographer who helped in shaping the mathematical format of the article. Blech's model was not included in this report, but Shechtman added Blech's name out of loyalty.

The new manuscript was sent to the prestigious periodical *Physical Review Letters*, where it was accepted for publication without delay and published on November 14, 1984. The article had a somewhat innocuous title: "Metallic Phase with Long Range Orientational Order and No Translational Symmetry."[8] The impact was tremendous, as if a floodgate had been lifted. Scientists, especially theoreticians, had been working on related problems, and those who were versed in the literature could easily connect their work to the Penrose pattern, Mackay's simulated diffraction diagram, and Shechtman's discovery.

Several weeks after the Shechtman paper was published, another paper—purely theoretical—appeared in the same journal, authored by Dov Levine and Paul Steinhardt of the University of Pennsylvania.[9] The Levine–Steinhardt manuscript was at the editor's office on November 2—less than two weeks before the Shechtman paper was published, but this was no coincidence. Levine and Steinhardt had learned about the Shechtman manuscript back in October from a Harvard professor to whom Cahn had sent a copy.[10] By that time, Levine and Steinhardt had produced a simulated diffraction pattern that resembled Shechtman's experimental pattern. Levine and Steinhardt knew they had to rush with publishing their theory because they realized that Shechtman's experimental discovery would generate a lot of interest.

The Levine-Steinhardt report had an elegant title: "Quasicrystals: A New Class of Ordered Structures," which then gave a name to the new structures that produced Shechtman's diffraction pattern. There are some who argue that giving a name to a discovery may be as important as making the discovery itself. A scientific discovery is made sooner or later; if not by this scientist, then by another scientist. In contrast, giving a discovery a name is an act that belongs to the individual who coined the name, and it is unique. Still, it cannot be denied that it is the discovery rather than its name that impacts science and the application of its fruits.

The discovery of quasicrystals was a big event in physics, crystallography, materials science, chemistry, and mathematics, and it even inspired artists to create sculptures and paintings reflecting these novel structures. The Penrose patterns have become popular decorations. The quasicrystal story shows the complex interrelationship of various branches of sciences and human activities. There were scientists like Alan Mackay, who were better prepared for Shechtman's discovery than Shechtman was, but, to Shechtman's credit, once he made his discovery, he grew to handling it in a seasoned way. He certainly wrote himself into the annals of the history of materials. Apart from his training in materials science, which included X-ray diffraction and electron microscopy, Shechtman's stubbornness and perseverance might have been his most important traits, enabling him to carry through with his discovery despite the tormenting loneliness of the many months of ridicule and rejection from the "experts."

Although large segments of the scientific community accepted the discovery of quasicrystals for a new form of materials, some very distinguished scientists held out for years. As Max Planck, who invented quantum theory—one of the most important discoveries in twentieth-century science—observed, "A new scientific truth does not triumph by convincing its opponents and making them see the light, but rather because its opponents eventually die, and a new generation grows up that is familiar with it."[11] This was literally the case with the quasicrystal discovery.

Shechtman experienced a great deal of frustration during the second half of the 1980s. The main antagonist to universal acceptance of the existence of quasicrystals was Linus Pauling, arguably one of the greatest chemists of the twentieth century. To the end of his life, Pauling maintained that Shechtman had observed twinned crystals, and that quasicrystals simply did not exist.[12] Pauling was a veteran of X-ray crystallography, but he did not much believe in electron crystallography, which is the method by which Shechtman did his experiments. Pauling's stand thwarted Shechtman more than anyone else's opinion because Pauling wielded such great authority. Even many people who would have accepted Shechtman's claim did not do so for fear of offending Pauling.

Pauling remained alert to new developments in science even when his primary interests had shifted elsewhere. When he heard about Shechtman's discovery, he asked Shechtman for more information, which Shechtman sent him. When Pauling suggested that Shechtman carry out additional experiments, Shechtman obliged and wrote up the results in a personal paper just for Pauling. Pauling did not have any problem with Shechtman's experiments, only with his interpretation of his observations. Eventually Shechtman visited Pauling and presented his findings in person. Pauling had many questions, but he could not be convinced.

The two continued to meet at various conferences; they were always cordial and always failed to come to an agreement. Shechtman's last encounter with Pauling was at a well-attended lecture, at which Pauling mentioned the quasicrystals and did so in a very negative way. Shechtman was in the audience, unrecognized by anybody around him. To Shechtman, Pauling resembled a politician and preacher rather than a scientist, someone who visibly enjoyed the admiration of the crowd. Pauling was doing his best to discredit the quasicrystal discovery and mentioned Shechtman by name without realizing that he was present. When Shechtman could not hold back any longer, he turned to his neighbors and said that

Pauling was wrong. The reaction of the people around Shechtman was such that he thought they might physically attack him.

A major turning point for the acceptance of nonperiodic structures came in 1987 when quasicrystals large enough for X-ray diffraction experiments became available. Shechtman presented the new findings at the conference of the International Union of Crystallography in Perth, Australia. The audience was convinced by the results and shortly thereafter formed a committee to redefine the term *crystal*, in order to include the materials that had previously been called *quasicrystals*.

People often refer to Shechtman as a "character" and as someone with the ability to polarize those around him. He can be flexible on many issues that he considers unimportant; in other areas, he has strong opinions and does not yield to peer pressure. He values his independence and does not take anything from anybody, making it a point to owe nothing to anyone who is not a close friend.

Once the discovery of quasicrystals was published in 1984, Shechtman's life changed. He became a celebrity almost instantly abroad and then gradually at home in Israel. First the nonscientific media discovered him, and then the newspaper *Haaretz* printed a front-page article about him and his discovery. He received speaking invitations in Israel, was elected to the Science Academy, and was awarded the highest prizes in his home country in addition to international recognition. There were times when he gave thirty annual invited lectures worldwide. In 1987, he became a full professor at the Technion and was appointed Philip Tobias Professor soon after.

The applications of quasicrystals have come about more slowly than initially expected. There are still hopes for applications in producing kitchenware coated with quasicrystals. Some see the material as outperforming Teflon, in that it is harder and does not scratch or peel off. Shechtman never participated in any patents related to his discovery.

The quasicrystal field soon grew larger than Shechtman could ever have imagined. It must be thrilling for him to see the growth of the field: the meetings, the books, and the research groups dealing with it all around the world. Soon after having published the discovery, Shechtman's major thrust of research shifted because he could not get funding for continuing his quasicrystals research. He may have been less persistent about raising money than sticking to his idea about his discovery. It may also be that he was looking for new adventures and new discoveries. After a while, though, Shechtman made a comeback to the quasicrystals field, a field in which he is now considered—not because of age but because of expertise and reputation—its great old man.

CHAPTER 9
RISKING REPUTATION
Conducting Polymers

I am a very lucky person and the harder I work the luckier I seem to be.
The late Alan MacDiarmid's
favorite Chinese proverb[1]

Alan MacDiarmid (1927–2007) was a well-known, established chemistry professor when a Japanese scientist showed him a most puzzling substance. It was an organic polymer, but it looked like a metal. MacDiarmid brought the Japanese scientist back to his laboratory in Philadelphia and initiated an intensive research program. The sponsor of MacDiarmid's research warned him about entering a field in which he had no experience and said that his reputation might suffer from a failure, though the sponsor continued its support. MacDiarmid wanted to enlist the cooperation of a physicist colleague, though that colleague declined, saying that MacDiarmid was risking his reputation by getting engaged in a dubious project. Fortunately, a fruitful relationship developed with another physicist. The team successfully created the conducting polymers, a family of highly useful substances with a plethora of everyday applications. The team of three, Alan MacDiarmid, Hideki Shirakawa, and Alan Heeger, shared the Nobel Prize in Chemistry in 2000.

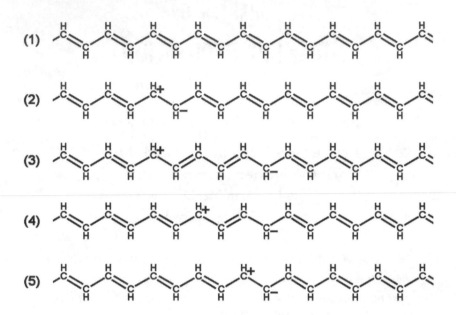

**Figure 9.1. Model of a conducting polymer: a segment of the polyacety-
lene chain with the moving electric charge along the chain. (1) Polyacetylene. (2) An electron is removed from one of the carbon atoms, and a
positive charge appears in its stead; the electron moves to an adjacent
carbon atom and there a negative charge appears. (3–5) These charges
move along the chain, and the chain becomes a conducting polymer.**

I n working on my thesis in 1965, I determined the structure of a
molecule in which a sulfur–nitrogen (S–N) bond played a central role. For years, after I had completed this piece of research, I
stayed alert for any additional information about chemical systems
in which such S–N bonds figured. In the early 1970s, I was invited
to give a seminar at the Chemistry Department of the University of
Pennsylvania. In addition to other strengths of this department, I
was aware of Alan MacDiarmid's early work on the crystal structure of $(SN)_4$ molecules; they were units consisting of four S–N

bonds forming a ring. During my visit, MacDiarmid told me about a polymer $(SN)_x$ in which the S–N bonds linked up in virtually infinite chains. A polymer consists of strongly bound basic units; thus, in $(SN)_x$ in which the basic unit is SN, the polymer is a chain of NSNSNSN, etc. MacDiarmid showed me the substance; it was gold-like, shining just as if it were a metal. I was impressed by this substance and even more by the enthusiasm with which he talked about its properties.

The next time I saw MacDiarmid, it was at the Nobel Centennial in Stockholm in December 2001. He was one of the many Nobel laureates who gathered on this special occasion in the Swedish capital, where I was invited to give a lecture on the Nobel Prize for the Royal Swedish Academy of Sciences. MacDiarmid had received the Nobel Prize one year before for his part in the discovery of conducting polymers. He told me that the festivities in 2001 were even more pleasant for him than those in 2000, because on this occasion he did not have to watch out for all the rules that a direct participant in the ceremonies must observe. He remembered my visit to his laboratory a quarter of a century before and my interest in S–N bonds. The sulfur–nitrogen polymer was the closest he had gotten to macromolecules before the discovery of polymers that conduct electricity; in short, conducting polymers.

The terms *polymers* and *macromolecules* are used interchangeably. There are many kinds of polymers, including the biopolymers such as proteins and nucleic acids, which are the carriers of life. It is ironic that during the first half of the twentieth century, many authoritative scientists adhered to the long-held view that macromolecules did not exist and that they all were colloidal systems; that is, aggregates of smaller molecules. In a way, the institution of the Nobel Prize issued its stamp of approval on the existence of macromolecules when one of the pioneers of macromolecular chemistry, Hermann Staudinger, was awarded the Nobel Prize in Chemistry in 1953.

In addition to the naturally occurring polymers, artificial macromolecules—often called plastics—had been in the forefront of modernization of both industry and everyday life from the early years of the twentieth century. When *Time* magazine compiled a list of the most significant contributors to science during the twentieth century, Leo H. Baekeland, the inventor of Bakelite, was among them. He was the only representative of the plastics field.

Baekeland was born in Belgium and received his education there; eventually he moved to the United States, where he announced the invention of a hard yet moldable plastic at a meeting of the American Chemical Society in New York in 1909. It was the first fully synthetic; that is, "human-made," plastic, which Baekeland created by combining two organic substances, phenol and formaldehyde. The appearance of Bakelite signified the beginning of the age of plastics. One of the most important aspects of Bakelite was that it made an excellent electrical insulation, which was one of its most widespread uses.

Further development of synthetic polymers occurred when a German scientist, Karl Ziegler, found a novel and very efficient method for polymerization using a new catalyst, which was an organoaluminum compound. Catalysts induce chemical reactions to take place under milder conditions than they would otherwise, or they make reactions possible that otherwise would not occur. The catalysts participate in the reactions, but in the end they reappear as they originally were, and are ready to be used again.

The Italian chemist Giulio Natta discovered that certain types of organoaluminum compounds can act as catalysts to synthesize so-called stereoregular polymers. These are substances in which the repeating units have uniform three-dimensional geometry. Nature has been known to produce such macromolecules, including cellulose and rubber, but Natta broke nature's monopoly over this area when he created such polymers artificially. Natta and Ziegler shared the Nobel Prize in Chemistry in 1963. Their polymers were still insulators, like Bakelite, in terms of their electrical properties.

The discovery of conducting polymers could not be pinpointed to a single "Eureka" moment, but there were some focal points of note. The Japanese polymer chemist Hideki Shirakawa was certainly present at their creation. He studied at the Tokyo Institute of Technology beginning in 1957 and completed his studies with his doctorate in 1966. He immediately began his career as a research associate in Professor Sakuji Ikeda's laboratory. Shirakawa acquired his first graduate student when Takeo Ito joined him at the start of Shirakawa's research career. They studied the conditions of forming polymers from acetylene, HCCH, which is a simple organic compound, a gas. Shirakawa and Ito did a lot of experimentation because by varying the experimental conditions, their products showed a broad variety of properties.

Basically, it was a simple experiment. They prepared a solution of the catalyst, using a fairly low concentration of these Ziegler-Natta substances, of which, again, there were a number of compositions. When the solution was ready, they bubbled the acetylene gas into the solution and continued stirring it to create the most homogeneous conditions for their mixture. The polyacetylenes produced had disappointing properties. They were black powders for which it was difficult to carry out even the simplest measurements of their properties, and they did not seem to be of great use, if any at all. This should not have come as a surprise to the team. Natta himself had produced polyacetylene, but neither he nor the other chemists were interested in this polymer because it did not show promising properties. Professor Ikeda was one of the few chemists who still wanted to learn more about this substance and especially about the mechanism of its formation.

There was a visiting scientist in Shirakawa's small research group by the name of Hyung Chick Pyun. He had already acquired his doctorate and was spending some time in Professor Ikeda's laboratory. Ikeda had charged Shirakawa with supervising the visitor's work. Shirakawa would give instructions to Dr. Pyun about the next experiment, and Pyun would carry out the experiment and report the outcome to Shirakawa. Dr. Pyun had grown up in Korea when the country was under Japanese occupation, and so he spoke fluent Japanese.[2] The fact that the scientists were able to communicate easily is important, because later reports suggest that the misunderstanding over the recipe for one of the particular experiments (which would earn well-deserved notoriety) originated from Dr. Pyun misinterpreting the Japanese instructions.

On a day that did not at first seem to stand out in any way from the team's usual routine, Shirakawa and Pyun had another of their usual discussions about the forthcoming experiments. Shirakawa told Pyun to have another go at the production of polyacetylene from acetylene, and he gave instructions as to the concentration of the catalyst to be used. Pyun prepared the reaction and, while constantly stirring, began to bubble the acetylene gas through the catalyst solution. As the product started forming, however, stirring became increasingly difficult, and eventually it became impossible. Rugged pieces of film were being formed in the reaction mixture. A peculiarity of the product was that it was shiny, giving the impression of a silvery metal, although Pyun knew it could not be a metal.

Pyun called Shirakawa. The two went through the various conditions of the experiment to determine what had gone wrong, and they figured out soon enough that Pyun had prepared the solution of the catalyst using a concentration one thousand times(!) higher than what Shirakawa had intended. Shirakawa could never decide if he had given the wrong instructions or if Pyun had misunderstood him. Had Shirakawa given the instructions in writing rather than orally, the mistake could have been avoided, and the discovery would not have been made.

All this happened in 1967, and although the team performed various measurements on the new substance, they did not find anything remarkable enough to warrant extensive further studies. But they published a few papers on the subject, the authors of which were the doctoral student, Ito; Shirakawa; and the professor, Ikeda.[3] Pyun's name did not appear among the list of authors even on the initial publication about the product of the fortuitous experiment. It is true that these articles were published years after the experiment, by which time the visiting Pyun had been long gone, and Shirakawa had lost contact with him. The whole story of this strange experiment might have disappeared into oblivion were it not for two circumstances.

One of these circumstances was that, for Shirakawa, the appearance of the silvery, metal-like substance in the polyacetylene experiment had remained remarkable, and so he had kept the sample at hand. The other was that in 1975, an American visitor, Alan Mac-Diarmid, gave a lecture at Shirakawa's workplace, the Tokyo Institute of Technology. The bulk of MacDiarmid's presentation was about molecular silicon compounds, but he could not miss yet

Figures 9.2a, 9.2b, 9.2c. From left: Alan Heeger in Stockholm, 2001; Alan MacDiarmid in Philadelphia, 2002; Hideki Shirakawa in Stockholm, 2001. Photos by the author.

another opportunity to show his shining gold-like $(SN)_x$ sample—the same he had shown me—which he carried in his pocket everywhere he went. After the lecture, Shirakawa and MacDiarmid talked, and Shirakawa showed MacDiarmid his own silvery sample of polyacetylene. The fact that this was an organic polymer was even more surprising than $(SN)_x$, and it set MacDiarmid's imagination in motion. Boldly, he invited Shirakawa to spend a year at his laboratory at the University of Pennsylvania (UPenn).

MacDiarmid had been at UPenn since 1955, having been born and brought up in New Zealand and having earned his first PhD at the University of Wisconsin and his second at Cambridge University, United Kingdom. He had become a respected inorganic chemist whose work received steady support from the Office of Naval Research (ONR). When he returned from his Japanese sojourn, MacDiarmid talked with his contact at ONR. It was a fruitful interaction, and he asked the ONR officer for approximately $23,000 to support Hideki Shirakawa as a postdoctoral fellow in his laboratory. The ONR man was a little reluctant, because polymer organic chemistry was not MacDiarmid's field; he was neither a polymer chemist nor an organic chemist. Nonetheless, based on the ONR's previous experience with MacDiarmid's research, his request was granted and Shirakawa arrived at UPenn in September 1976 for a one-year visit.

Shirakawa had not been particularly interested in electric conductivity of polyacetylene, and when he measured it, it did not show much conductivity anyway. The first task to which MacDiarmid and Shirakawa addressed themselves was to see whether the electric conductivity of their polyacetylene samples could be enhanced. The focus on this property was justified by the metallic appearance of the substance and by MacDiarmid's prior experience with the sulfur–nitrogen polymer. As chemists like to work with pure samples, the team set themselves to purify their polyacetylene.

To their surprise, the purer the sample became, the more its electric conductivity diminished. This gave them the idea of purposely making less pure samples. The procedure of contaminating on purpose is called *doping*; this procedure has become the distinguishing feature between conducting polymers and all other polymers.

The doping increased the electric conductivity of the polyacetylene samples ten million times! The essence of the process is that the very rigid electronic structure of alternating single bonds and double bonds in polyacetylene is loosened by the addition of a few percent of agents; for example, halogens. These attract electrons away from the chain, thus creating electron holes (the locations of the missing electrons); and these emerging electron holes are easy to transmit along the polymer chain; hence the conductivity.

MacDiarmid understood soon enough that in addition to polymer chemistry and organic chemistry, yet another domain of science was being involved in these conducting polymers—condensed-state physics, the science of solids and liquids. He therefore invited a physicist from the excellent neighboring Physics Department of UPenn to join their quest for creating this heretofore unheard-of new field of materials.

Soon Alan Heeger, a UPenn physicist, became the third member of MacDiarmid and Shirakawa's winning team. The story of their initial contact is worth retelling because it shows that even specialists of fields as closely related as inorganic chemistry and condensed-state physics need to be concerned about speaking a common language. According to Alan Heeger, it was not long before MacDiarmid was to leave for Japan that he went to MacDiarmid's office and heard his description of "SNX," a wonderful material.[4] MacDiarmid told him that it had metallic properties and suggested a joint study, but Heeger was not interested. The reason for Heeger's lack of interest was that when MacDiarmid referred to poly(sulfurnitride), $(SN)_x$, Heeger, a condensed-state physicist, assumed that MacDiarmid was talking about Sn_x; that is, metallic tin (Sn, for the Latin *stannum*, is the chemical symbol of tin.). The pronunciation of the two formulas, $(SN)_x$ and Sn_x is the same, and the misunderstanding was resolved once the men scribbled down the two formulas. From that point on, they built up a wonderful cooperative relationship that led to spectacular results in their work on conductive polymers. But it almost did not happen.

As alluded to above, MacDiarmid recognized the need for involving a physicist in the conducting polymers project right at the beginning of his work with Shirakawa. However, Heeger was not his first choice.[5] Before asking Heeger, he had asked another member of the physics department, but he would not reveal this other person's name when he was telling me this story. He told this other physicist about what he and Shirakawa had done and what effects they had found, including the enormous increase in conductivity upon doping. MacDiarmid suggested joint work over the interpretation of the phenomenon. He was then gravely disappointed when his physicist colleague said, "Alan, this is just a junk effect." His advice to MacDiarmid was: "Don't touch it." He thought MacDiarmid would ruin his hard-earned reputation, and, needless to say, he declined to join the team. In contrast, Heeger was willing to take a risk.

Many years elapsed from the time of Pyun's serendipitous experiment with the exceptionally high concentration of catalyst solution to the first publications about the production and properties of their doped polyacetylene during the second half of the 1970s. But it was with lightning speed that, within weeks after Shirakawa's arrival in Philadelphia, a pivotal observation occurred. The team conducted numerous experiments, but rather than merely observing what was going on, they manipulated the results by testing many different experimental conditions. Eventually, they reached a "Eureka" moment: the breakthrough was a bromine doping experiment on Tuesday, November 23, 1976.[6]

Electrical conductivity of the polymers was a fundamental discovery, and it suggested the possibility of numerous applications. Other properties also showed marked changes under the conditions of varying degrees of doping; for example, the optical properties. Initially, the materials were neither sufficiently stable nor processable, so they had to be improved before serious applications could be contemplated. A lot of chemistry came into play during this process, because changing the composition of the polymers changed their properties—but the team wanted only such changes in the properties that would not hurt the basic ones that made their use advantageous in the first place. They soldiered on, and by sometime in the 1980s—according to Heeger—the materials became more mature.[7]

As soon as the materials became applicable, physics was employed for their further development, since the best way to appreciate the uniqueness of the conducting polymers was by comparing them to metals. Conducting polymers essentially have all the properties that metals have, and for which metals are so widely used, but these polymers have some additional, valuable properties that metals cannot match. The conducting polymers can be kept in solution, in a bottle; that is, dissolved in a solvent. When the solvent is evaporated, the conducting polymer remains. No metal can do that. Thus, for example, polyaniline can be made into a conducting polymer; it can then be dissolved in toluene. When the toluene is allowed to evaporate, the metal-like substance is regained. Such solutions can be used as inks, which can be printed. A typical application is for solar cells: large areas can be printed at a low cost. Light-emitting devices and diodes can be made using the conducting polymers; this is just one example of such applications in electronics.

Conducting polymers may not be as good conductors as some of the best metal conductors, such as copper, but they may have other valuable properties. The conducting polymers are very sensitive to various chemical interactions beyond doping. Thus, for example, the electric conductivity of polyaniline decreases by six (!) orders of magnitude when ammonia gas surrounds it; hence it is an excellent device for acting as an ammonia sensor. Generally speaking, conducting polymers show variability in conducting and in other properties, too, for which there is no match among the metals.

MacDiarmid and Heeger did not start filing for patents right away, as their first inventions appeared in connection with conducting polymers. When Heeger and his colleagues started making diodes by casting a film from solution, it seemed so simple that they decided not to patent it—to their later regret. Eventually they patented their inventions, as well as their improvements on other people's inventions. From their small circle of researchers, a huge field has emerged. Heeger's opportunities broadened when he moved from UPenn to the University of California at Santa Barbara in 1982.

The international conferences on conducting polymers attracted a few hundred people in the early 1980s, and thousands in the 1990s. Companies mushroomed, including Heeger's own UNIAX, which he and his partners sold in 2000, shortly before receiving the Nobel Prize. Heeger told me that if they had waited until after the Nobel Prize announcement, they could have made a better deal. But who knew if and when the prize would come, though Heeger started thinking about the possibility in the early 1990s.

Heeger, Shirakawa, and MacDiarmid shared the Nobel Prize, and it seems that all three were needed to bring out the discovery and make it successful. It was fortunate that the rules of the Nobel Prize allow three persons to share one prize in a given category, so in this case nobody was left out. However, we should not forget the pivotal role of the visiting Korean scientist, Dr. Pyun, in causing the events that led to the discovery. The first preparation came about by his hand, even if he was acting on Shirakawa's instructions, which he appears to have misunderstood. This in itself is quite curious. Pyun had already earned his doctorate by that time, so he must have been well versed in chemical synthesis. Catalysts usually come in rather low concentrations, so Pyun should have been at least somewhat surprised that Shirakawa would want him to apply an enormously more concentrated solution than previously. Why did he not ask Shirakawa about it when he received his instructions? This is quite puzzling.

We can look at this question from at least two angles. One is that Pyun was just following instructions without giving them much thought. It did not seem strange to him that the catalyst was in such a high concentration. Had the team been experimenting with reactions in which they used catalysts in a wide range of concentrations, this anomaly might have made sense. However, the fact that Pyun did not ask about such an enormous amount of catalyst that seemed to appear out of the blue might indicate that he was following instructions mechanically. Furthermore, when he produced a thick product that was so dense the mixing rod could hardly be moved, he purportedly did not try to understand why this happened. He might have tried to figure out what had gone wrong; at this stage it might have occurred to him that something in the recipe was wrong or at least strange. Instead, he reportedly just went back to Shirakawa to show him what he had obtained according to Shirakawa's instructions.

If we look at the story from an entirely different angle, it could be assumed that the visiting scientist simply produced a new substance. He did something nobody had done before, and although he did it by mistakenly interpreting somebody else's instructions, the mistake and the consequences were his alone. Of course, it is impossible to know whether other people in other laboratories and at other times might have produced the same substance or other, similar substances, and whether those substances might have been as quickly discarded as products obtained by mistake. A classic example of this is the case of Alexander Fleming, who observed that a mold was consuming a bacterial culture that he was culti-vating for an experiment. He might have thrown out the Petri dish and begun producing his bacterial culture anew. Instead, he forgot about his original intentions and decided to try to understand why the mold would consume his bacterial culture. From this point on, the road was open to the discovery of penicillin (though it was not as straightforward as it is often imagined).

It was Shirakawa who found what happened to be strange and decided to try to understand it. It was because of his curiosity that the misunderstanding was cleared up. What caught Shirakawa's attention was the blistering metallic luster he noticed. And when Shirakawa later attended MacDiarmid's lecture in which the Amer-ican displayed the lustrous gold or copper-like sulfur–nitrogen polymer, he showed MacDiarmid his own silvery, metal-looking hydrocarbon polymer. From this point on, things started happening. Shirakawa did not pursue the conductivity of his substance with the metallic appearance. He treated the substance more as a curiosity than as something that should be further explored, which is why it would be unfair to accuse the Americans of going to Japan, noticing an important discovery, and then expropriating it for their own goals. The truth is that, without their involvement, there would have been no discovery, period.

Had MacDiarmid not demonstrated his inorganic polymer during his lecture, at which Shirakawa was in attendance, nothing might have happened. But it could also be said that had the visiting Korean scientist Hyung Chick Pyun not produced the curious polymer, nothing might have happened, at least not for a period of time, whether days, months, years, or decades, we do not know. Sooner or later, though, there could have been another experiment in another laboratory, or calculations might have later arrived at suggesting a conducting hydrocarbon polymer. Scientific discoveries are not like artistic creations that never happen if the particular artist does not create them.

Having said all the above, it is a pity that Dr. Pyun has disappeared from view; it would be interesting to have his take on the matter (and I have tried to find him, with no success). The paper Shirakawa published with his professor and coworker did not include Pyun among its authors; Shirakawa did not mention his name in the main part of his Nobel lecture, not even when he described the "fortuitous error in 1967,"[8] but he did refer to Pyun in his acknowledgments at the end of his Nobel lecture as "with whom [Shirakawa] encountered the discovery of polyacetylene film by the fortuitous error."[9]

Shirakawa's merit was in having had the polyacetylene produced and in noticing its metallic appearance. He also had the courage to show it to MacDiarmid after MacDiarmid's lecture. Then, of course, when MacDiarmid brought him to Philadelphia, he worked hard for the success of the substances, but this was not such an original contribution.

MacDiarmid had always had an interest in metal-like polymers, but they were inorganic polymers. He was a chemist who liked to play with substances—a very positive trait that contributed to this discovery. However, had the Japanese scientist not produced the strange substance, it is unlikely that MacDiarmid would have stumbled into the conducting organic polymers. His main merit was in grabbing the opportunity presented to him by fate in the form of Shirakawa's approaching him. Without hesitation, he invited Shirakawa to his laboratory even before he had secured the necessary funding. By doing so, he took a risk, and he took another kind of risk when he ventured into a field in which he had no experience. He received and ignored repeated warnings that he was risking his hard-earned reputation, and just charged ahead. He was ready for the discovery, and he was ready to take a risk for it. Had he not done so, he would have completed his career in a very respectable, but not extraordinary, way.

MacDiarmid was a person of enormous enthusiasm and energy. When he received his Nobel Prize, he was well over seventy years old. He was not in the best of health, yet he was busy as ever. He continued doing research and directed the work of his associates while coping with the obligations of a Nobel laureate. When he had problems scheduling a discussion with a German postdoctoral fellow in his laboratory, due to the excruciating demands on his time that stemmed from his sudden fame, his coworker grudgingly remarked: "This Nobel Prize came at a most unfortunate time."[10] Indeed, the Nobel Prize came a little too late for MacDiarmid, because he could not take full advantage of the enhanced possibilities that the award would have provided him. He passed away at the peak of his plans and ambitions when his failing health could not keep pace.

Alan Heeger is a physicist with an interest in the interphase between physics and chemistry. He, too, would have had a respectable, but not extraordinary, career, had he not gotten involved with MacDiarmid's project. Once he did, he took up the lion's share of work in bringing the conducting polymers to triumph. However, his involvement almost did not happen. Early on, MacDiarmid very correctly recognized that he would need the involvement of an expert in electronic structures in the condensed phase, which is rather far removed from the immediate interest of most chemists, even of those who are interested in the electronic structures of molecules. But we have seen that MacDiarmid had turned to Heeger only after another physicist declined his invitation. In contrast with MacDiarmid's first choice, Heeger liked to take risks and was open to MacDiarmid's proposal.

In hindsight, it is difficult to determine how busy Heeger would have been had he not switched his research to the conducting polymers; nor can we know how extensive that switch might have been. Had he been working on an exciting project, he might have found it difficult to switch to something else, but even that is impossible to know. In any case, without MacDiarmid, Heeger would not have gotten involved in the conductive polymer project. It was his entrepreneurship that helped the project triumph.

Considering the roles of the three-man team, MacDiarmid was the key person in the conductive polymer discovery. He launched the project, and he was the one who brought in the other two scientists. In the further development of the conducting polymers, from scientific curiosity to broad-based and versatile applications, Heeger's involvement was instrumental. Perhaps the decade difference in MacDiarmid's and Heeger's ages, as well as Heeger's excellent entrepreneurial skills, played a role in their diverging activities. Entrepreneurship was especially appropriate due to the nature of the discovery in that it was fit for numerous applications.

Shirakawa was also a decade younger than MacDiarmid, yet he had retired months before the announcement of the Nobel Prize, not only from teaching but also from research. What is more, the enhanced opportunities a Nobel Prize accords its recipient did not lure Shirakawa back to his prior activities, nor did it catapult him to new ones. Shirakawa retired in 2000 after a solid but not extraordinary career. He had moved to Tsukuba University in 1979—a distinction in itself, as Tsukuba is very research oriented. But the only other honors he received were two Japanese awards in polymer science. After receiving the Nobel Prize, he was awarded the two highest Japanese recognitions, both within weeks of the Nobel Prize announcement.

The Nobel Prize invigorated Heeger's and MacDiarmid's work. Neither was a member of the National Academy of Sciences at the time of the Nobel Prize, but they were elected soon after—Heeger in 2001, and MacDiarmid in 2002. Other awards, honorary doctorates, and all manner of distinctions followed, but each had received recognitions before the Nobel Prize. Heeger has been active and productive ever since joining the project of conducting polymers. MacDiarmid's determination to continue his work and activities was admirable, particularly in light of his illness. He passed away just as he was preparing for another visit to his native New Zealand. It was a sad loss when he died at the height of enjoying his life and work, but it was perhaps his work that added enjoyable time to his life.

CHAPTER 10
RELUCTANT ENVIRONMENTALIST
Saving the Ozone Layer

. . . the calculations indicated that if we go on releasing the CFCs, the amount of ozone is going to be severely depleted.
F. Sherwood Rowland[1]

F. Sherwood Rowland (1927–) did not start out as a crusader for the environment. His first environment-related research exonerated industry from accusations of mercury poisoning of the fish in the oceans. His next research project, however, led to the discovery in 1973 that the millions of tons of CFCs used for a variety of purposes and released into the atmosphere might destroy the earth's protective ozone layer in the stratosphere. Once Rowland realized the serious implications of his findings, he became a vocal advocate dedicated to stopping the process that would have led to damaging life on earth on a major scale. A long struggle ensued for an international ban on producing and marketing these substances. The battle culminated in the creation of an international treaty on environmental protection in 1987 with the signing of the Montreal Protocol on Substances That Deplete the Ozone Layer.

Sherwood Rowland grew up in Delaware, Ohio, and graduated from high school one month before his sixteenth

Figure 10.1. F. Sherwood Rowland and Mario J. Molina at the University of California, Irvine, 1975. Courtesy of Special Collections and Archives, University of California, Irvine Libraries. University Communications Photographs. AS-061. Box 45: A75-162, image [10A].

birthday. His science teacher entrusted him with operating the local volunteer weather station, but other than that, he was hardly interested in issues related to the environment. Rowland enrolled at Ohio Wesleyan University in his home town. But when World War II was raging, he joined a navy program at eighteen. He finally graduated from university in 1948.

He went on to graduate school at the University of Chicago, where the roster of his teachers in physics and chemistry was spectacular; most had come from working on the Manhattan Project. Rowland took courses from the Nobel laureates Harold Urey and Enrico Fermi and from future Nobel laureates Henry Taube and Maria Goeppert Mayer. The no less celebrated Edward Teller was also among his professors. Willard Libby, also a future Nobel laureate, was Rowland's mentor for his doctoral work. Libby discovered archeological dating with carbon-14 isotope, and he impressed upon Rowland the usefulness of radioactive isotopes in research.

Libby taught Rowland to have fun and to find excitement in science, and that such joy and excitement come from doing new things. For Rowland, it was a life lesson to distinguish between being "in the groove" and being "in a rut." The first means that things are going well, and the second that the researcher is trapped into doing the same thing over and over again. The boundary is not very clear when it comes to getting into the groove of truly understanding what one is doing, on the one hand, and then feeling that the process is becoming routine, on the other. Genuine researchers feel that if they know what the next experiment will result in, they might as well not do that experiment. Having this philosophy in mind, Rowland made a conscious effort to introduce a new direction in his research every six to ten years, though the application of radioactive isotopes remained a common thread in his projects. In addition to Libby's influence, Rowland received steady research support from the Atomic Energy Commission (AEC).

After earning his PhD degree in 1952, Rowland went to Princeton University for a few years. There, the department chair frowned upon Rowland's having received an offer of AEC support and warned him that he was too young to have independent funding. Rowland decided to move on. He was offered a tenure-track faculty appointment at the University of Kansas in 1956 with plenty of space and full independence. Eight years later, in 1964, he joined the University of California, Irvine, as full professor, and as of this writing he's still there. The move to Irvine was another challenge: Rowland was the founding chair of its new Department of Chemistry.

Rowland was a respected professor in the areas of education and research, but nothing pointed to an extraordinary career during the next few years. In the late 1960s, he felt, once again, that it would be desirable to change his research direction. He initiated some experiments involving radioactive chlorine and radioactive fluorine for which he needed an inert source to hold the stable isotopes. This is how Rowland came across chlorofluorocarbons (CFCs) for the first time: the compounds consisting of carbon, chlorine, and fluorine.

Still looking for new research direction, in 1970 Rowland attended a meeting in Salzburg, Austria, on the environmental applications of radioactivity. This was his first encounter with environmental issues in his work. The meeting was organized by the International Atomic Energy Agency (IAEA), which was very much interested in the applications of isotopes and wanted to learn more about radioactive fallout in the atmosphere from nuclear testing. Rowland returned home from the meeting with an idea for a new project related to a grave environmental concern at the time. There was fear that mercury was polluting the fish that people ate, and Rowland wanted to collect data on the mercury content in fish as a function of time. Accordingly, he was looking for fish having been caught at different times.

Examining swordfish—a marine fish with an extended sword-like upper jaw, sometimes called a bill—interested Rowland, because he thought the mercury might be concentrated in the bill as metallic Hg^{2+} ions. Seeking to establish a trend in mercury content in fish, Rowland and his associates carried out their first experiment with a recently caught swordfish. Next, they knew they would have to hunt down a preserved swordfish for the following experiment. Rowland supposed that the bills, if not the whole swordfish, might be preserved as hunters preserve trophies. The idea of mercury accumulating in the bill of the swordfish did not turn out to be valid, however, because the mercury accumulates in the flesh, presumably present in organic compounds that contain methyl mercury.

A San Francisco museum displayed a preserved swordfish that was being saved for its oddity rather than for its being a swordfish.

It must have had a terrible accident when it was small, as it looked like it had stabbed itself and grew up with its sword bent back and buried in its forehead. Due to this accident, it was not able to fish on its own and so must have survived on leftovers: swordfish do not swim individually but in pods. This swordfish did not appear to be as well fed as most swordfish are, and so Rowland was not sure whether the mercury content of the food intake of this disabled swordfish was characteristic of a normal swordfish. Another source of uncertainty was whether preservation techniques involving drying and other approaches would impact the mercury content.

Rowland and his associates carefully analyzed this particular swordfish, which was caught in 1946 off the coast of San Diego. At the time of the analysis, in 1971, it provided data of a twenty-five-year-old sample, which was valuable, even if it was only a single piece of data. The team obtained further information from some samples of pipe fish, tiny organisms that were preserved at the same time as the odd swordfish. The gist of the analyses was that the mercury concentration in the twenty-five-year-old swordfish was essentially the same as that recorded for freshly caught swordfish.

Furthermore, the team managed to get some samples of tuna fish saved at the Smithsonian Institution in Washington, DC—the oldest being from 1878—that showed, again, the same levels of mercury as fresh tuna. The resulting publication in *Science* documented the approximately steady level of mercury in fish over a century.[2] Thus, the presence of mercury in oceangoing fish was not a consequence of industrial pollution; the mercury must have worked its way through the food chain with no outside interference from the modern world. The conclusion was that tuna and swordfish contain dangerous levels of mercury and represent an environmental problem—but not as a result of industrial pollution.[3]

Rowland at that time was not very interested in pollution, hazardous materials, or other environmental issues, but he could not insulate himself from them either. The first Earth Day on April 22, 1970, was a big event, one in which Rowland's family members were involved. Their interest rubbed off on Rowland, who, as always, was constantly alert for potential research projects. At the time of the first Earth Day, however, the pollution of the atmosphere was not yet a matter of wide concern. The atmosphere was assumed to be limitless (except for the smog in places like London and Los Angeles at the time).

When the British scientist James E. Lovelock published his pioneering observation in 1971 about the presence of fluorine compounds in the atmosphere, he did not consider them to be a hazard either; rather, he found them useful for tracing the air movement over the oceans and continents.[4] Lovelock then started measuring chlorofluorocarbons, and even as late as 1973, he stressed that their presence in the atmosphere constituted no hazard and that they were useful as inert tracers.[5] The chlorofluorocarbons were unique due to their inertness. Many human activities release naturally halogenated molecules, but the CFCs are exclusively synthesized by humankind. Rowland first heard about Lovelock's measurements of CFCs at an AEC-sponsored scientific meeting in 1972.

Rowland's growing interest in applying his radioactive isotope techniques to studying environmental problems and thinking about CFCs gave him the idea to formulate a new research project. He undertook this project with no clue about where it might lead, and he certainly did not have an agenda that would relate his project to the environmentalist movement. When he was preparing his annual proposal for the AEC in 1973, he added a minor item to the thrust of his proposal, one that did not involve radioactivity. It was about branching out and looking at the fate of the CFCs in the atmosphere. His candidate for this work, Mario Molina, had just joined his group as a postdoctoral fellow. Rowland had a routine arrangement in his laboratory. As principal investigator, he made the proposal, and when the AEC agreed to it and allocated the money, Rowland employed someone for the project. That is how Molina became involved in this investigation.[6]

The term *CFCs* has taken on negative connotations but only due to their utilization in the way they were let into the atmosphere. Even today they are being produced in large quantities and serve as precursors for Teflon. CFCs were produced for the first time by an obscure Belgian scientist, Frédéric Swarts. In the late 1920s, American chemist Thomas Midgley Jr. improved the process of obtaining CFCs with the aim of replacing toxic refrigerants such as ammonia, chloromethane, and sulfur dioxide. Midgley's substances satisfied all the requirements prescribed for the desired refrigerants: they were volatile, nontoxic, and nonreactive. Midgley demonstrated their properties in 1930 when he inhaled the gas form of his CFCs and then blew out a candle with it.[7] In time, CFCs were used in a variety of ways, serving as refrigerants, fire extinguishers, propellant in aerosol cans, and more. They were used in military aircraft during World War II and later in warships, mostly as firefighting materials. Midgley became a celebrated scientist in his time; the American Chemical Society awarded him its most prestigious award, the Priestley Medal, and the National Academy of Sciences elected him a member. The CFCs quickly became a success story, but there was still much unknown about them.

It is an interesting question what happens to chemicals when they are discharged into the atmosphere. There may be a variety of fates, but not a large variety; there are three possible scenarios. Some molecules absorb light and decompose under its impact. For example, the green chlorine gas consisting of chlorine molecules Cl_2 absorbs sunlight and decomposes into chlorine atoms. Hydrogen chloride, HCl, is different; it is transparent, and light does not affect it. On the other hand, it dissolves in water, so it remains in the atmosphere only until it rains, at which time it returns to the ground with rainwater. Other substances may stay in the atmosphere longer if they do not disintegrate in light or dissolve in water. Such substances must wait until a reactive radical—a result of disintegration of another substance—appears, which oxidizes them. This may take months or even years, but eventually these substances also return to the ground.

CFCs are different from other chemicals in that they are so inert that none of the three scenarios described above applies to them. Nothing happens to them in the lower atmosphere—the troposphere—which extends up to about ten miles above the earth's surface. The estimated average lifetime of CFCs may be measured in tens of years or even longer. Although CFCs are heavier than air, winds pick them up, and they slowly reach the stratosphere, which is in the region between ten and thirty miles of altitude. There, the destruction of CFCs may be caused by sunlight, which has higher energy than in the lower atmosphere. Sunlight is composed of radiation of various wavelengths (and, accordingly, of different energies): red to violet, plus infrared and ultraviolet. In the upper atmosphere, there is more ultraviolet, hence higher energy. Whereas the temperature rapidly cools at the higher altitude of the troposphere, yet higher, in the stratosphere, there is a warming again, and at thirty miles of altitude the temperature may be around twenty-five degrees Fahrenheit. This is not warm, but it is certainly warmer than the higher layers in the troposphere. A peculiarity of the stratosphere is that there is an ozone layer of sparse density, but it is very thick. It spreads in the region between fifteen and thirty-five miles of altitude.

An ordinary oxygen molecule consists of two oxygen atoms linked to each other. Because it is a very stable molecule, it is not broken up by sunlight reaching the surface of the earth. However, in the stratosphere, sunlight has higher energy and can break up the oxygen molecule; the resulting individual oxygen atoms then may each join another oxygen molecule. Such a union results in an ozone

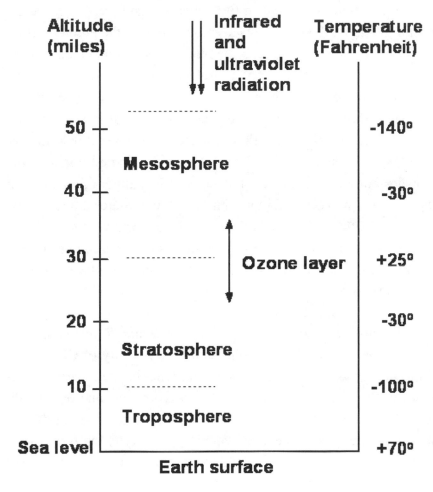

Figure 10.2. Schematic representation of
the structure of the earth's atmosphere.

molecule, which consists of three oxygen atoms forming a triangular
shape, in which each of the oxygen atoms occupies an apex of the
triangle. Ozone was discovered in 1839 by C. F. Schönbein, who
named the substance after the Greek word *ozein*, "to smell," in reference to its characteristic odor. Lightning or electrical discharges
cause this odor due to some ozone production. The ozone molecule
is much less stable than the ordinary oxygen molecule.

Some ozone is present in the lower atmosphere, but it has no useful role there. It contributes to smog as part of the pollution from motor vehicles and is toxic to the human respiratory tract. In contrast, ozone in the stratosphere protects life on earth. It absorbs much of the ultraviolet light from the sun that would be harmful to us. The ultraviolet radiation from the sun can cause sunburn and skin cancer when the protection from the ozone layer is insufficient. The US Environmental Protection Agency estimated that a 1 percent drop in global ozone could increase the number of annual skin cancer cases by twenty thousand in the United States alone.[8] Actually, by absorbing the ultraviolet light, the ozone layer produces the relatively high temperature of the stratosphere, creating another kind of protection for the earth; namely, confining the weather to the troposphere.

Rowland and Molina decided to perform a series of thought experiments and calculations. They collected and utilized all available information in the literature about the stability of CFCs, how molecules react in the atmosphere, and what happens when molecules in the atmosphere receive energy from sunshine or rainfall. They soon came to the conclusion that due to their volatility and inertness, the CFC molecules would eventually reach the stratosphere. Only there would the ultraviolet light from the sun be able to break them. As a result, there would be chlorine atoms, which are very reactive and which would be able to react with the ozone molecules, which are not very stable in the first place. The reaction is simple to depict:

$$Cl + O_3 \rightarrow ClO + O_2$$

In this reaction, the chlorine atom destroys the ozone molecule and forms chlorine monoxide, ClO, and normal oxygen. In addition, there would also be some oxygen molecules, because ozone is not very stable and may disintegrate into a normal oxygen molecule and an oxygen atom. If there is nothing for the oxygen atom to react with, except the normal oxygen molecule, chances are good that the two will recombine into an ozone molecule. However, if an oxygen atom collides with the chlorine monoxide molecule, again, a reactive chlorine atom emerges as well as a stable oxygen molecule:

$$ClO + O \rightarrow Cl + O_2$$

The bottom line is that the reactive chlorine is being regained as a result of these two reactions, whereas one oxygen atom and one ozone molecule yield two normal oxygen molecules in yet another reaction leading to the destruction of yet another ozone molecule:

$$O + O_3 \rightarrow 2O_2$$

If Rowland and Molina had the means to travel to the stratosphere and carry out measurements there, they would have been more easily convinced that their calculations were correct. But this was not possible at that time, so, lacking the ability to verify their findings through experimentation, they could rely only on calculations. First, they had to convince themselves that the CFCs posed an unprecedented danger to the ozone layer. They came to this conclusion shortly after they had submitted their proposal to the AEC about their research plan to investigate the fate of CFCs in the atmosphere.

In the fall of 1973, when Rowland and Molina's calculations were being carried out, the annual production of CFCs was close to one million tons. Even if the team assumed that production would not increase, their estimate of the loss of the ozone layer was at about 7 to 13 percent. It was reasonable to suppose, however, that production of CFCs would be rising; by using a conservative estimate of an annual 10 percent growth, the expected loss of the ozone layer appeared much greater than if the CFC production were to stay constant. It was especially frightening that once a free chlorine atom was produced, it initiated a chain reaction, which again and again re-created the free chlorine atom. The two researchers realized that they were facing an ozone removal process that had the potential of becoming the dominant process in the stratosphere. The project that started as an interesting scientific exercise was fast turning into a wake-up call about a potential environmental disaster.

What Rowland and Molina saw in their numbers was so shocking that they first thought they must have made a mistake in their calculations. The error should have been not just a factor of two or three or even ten but a factor of a thousand! However, the calculations were not that complicated, and whatever way they checked them, the results did not change. By December 1973, Rowland and Molina knew they had uncovered an environmental problem of global significance. Their initial hesitation about the validity of their findings came partly from the enormity of the effects and partly from the fact that their calculations and consider-ations were so simple that it seemed surprising that no one before them had come to similar conclusions. Such hesitation is quite characteristic when a researcher makes a discovery, especially when it seems—at least in retrospect—simple.

There were several other scientists who had been involved in studying the atmosphere and the impact on it of various chemicals—both natural and human-made—that went into the atmosphere. And there had been notable clashes with various industries, and not only with the chemical industry. There was, for example, the environmen-talists' fight against the high-flying supersonic jets releasing nitrogen oxides that could cause depletion of the ozone layer. This fight was based on Harold Johnston's research at the University of California at Berkeley, the results of which played a decisive role in the US

Senate's decision to overturn a bill that would have made it possible to build supersonic jets for civilian transport in the United States—like the British-French Concorde and the Soviet TU 144.

One of the best-known international scientists in this field is Paul Crutzen, who was working in Stockholm at the time. He was primarily concerned with the harm nitrogen oxides might cause to the atmosphere. Other scientists, notably Richard Stolarski and Ralph Cicerone at the University of Michigan, were concerned with the hydrogen chloride emissions from the solid-fuel rockets used for the space shuttle by NASA. Rowland and Molina's project would bring more scrutiny to the potential damage of CFCs. Chlorine's impact on the ozone layer had been investigated before Rowland and Molina had begun their work, but the chlorine sources—whether natural or human-made—were not big enough to cause a splash, compared to the potential consequences of CFCs.

Rowland and Molina visited Johnston at the end of December 1973, informing him of their findings and showing him their calculations. Johnston encouraged them to make their results public, but he warned them to be prepared for the consequences. Rowland and Molina went into immediate action and readied a paper for publication in *Nature*. The magazine received their manuscript on January 21, 1974, but due to some mishaps, the paper was not printed until five months later, on June 28, 1974. *Nature* often publishes important communications within weeks of acceptance, but this was not the case with Rowland and Molina's manuscript. Nevertheless, once *Nature* finally published the article, the discovery was out in the open—still, Rowland and Molina had to be disappointed, because it generated hardly any interest. The road to a broad recognition of their discovery was a long one, and when it finally happened, the negative reaction outweighed the appreciation for quite some time.

By the time the *Nature* paper appeared, Rowland and Molina knew much more about the whole issue, and the question then was how to get out the information in more detail. To publish a long paper in a peer-reviewed journal could take many months. Instead, Rowland labeled their detailed account as an AEC report and distributed two hundred copies. Eventually, it also appeared in a journal in a slightly revised version.[9] In addition, Rowland was among the participants of the American Chemical Society's press conference in September 1974, where he had an opportunity to publicize his and Molina's discovery. The Associated Press compiled the story and distributed it to four hundred newspapers with a total circulation of one hundred million people, but the team's findings were still far from making a splash.

The story received national exposure when it appeared on the front page of the *New York Times* in September 1974.[10] The story was then published in *Time* magazine and was featured on the *CBS Evening News* with Walter Cronkite. One might have expected that from this point on, Rowland and Molina would have easily gained more and more recognition, but this was not the case. The period between 1974 and 1988 was rather bumpy for Rowland. He became a controversial figure because he was at odds with much of the chemical industry. He was already an established scholar, so he could sense the difference in attitude toward him more easily than a young postdoctoral fellow—as Molina was at the time—would have been able to do.

Rowland was used to getting speaking invitations, but during this period he received hardly any from chemistry departments, though he was asked to give talks at other university departments such as toxicology, geology, and physics. Rowland compared his situation with the American Western films, in which the good cowboys wore white hats and the bad guys wore black hats. He felt that his symbolic hat was portrayed as black as it could be.[11] He and Molina used to read the monthly publication *Aerosol Age*, in which they found the most bizarre accusations against them. Once, according to Rowland, they were labeled as KGB agents, intent on disrupting American industry.[12]

Even James Lovelock, who had pioneered work on the observation of the ubiquity of CFCs in the atmosphere, accused the American team of being prone to panic, saying that Rowland in particular was acting like a missionary. Then Lovelock, to give emphasis to his accusations, added, "The Americans banned tuna fish and they blamed industry until someone went to a museum and found a tuna fish from the last century with the same amount of methyl mercury in it."[13] Lovelock did not know that that "someone" was Rowland himself. It was a lucky break for Rowland, because Lovelock's story enhanced rather than diminished Rowland's credibility. It is worth noting that since then Lovelock has reversed his views on the dangers to the environment in general, and on the American approach to them in particular. He was the one who came up with the idea that the earth should be viewed as a living organism, giving it the name *Gaia* after the Greek goddess of the earth.[14]

For years, Rowland attracted very few American postdoctoral associates, and so his research was carried out by foreign postdocs, mostly from Japan. Rowland's peers in the United States sensed that his area of research was suspicious or even controversial, and he was further ridiculed by the industry. As a result, graduate students were advised not to get involved with Rowland; this situation lasted about fifteen long years. Fortunately, even though he seemed to be an outcast in his profession, Rowland's research support from the AEC continued steadily, and his research flourished during this period. The difficulties he experienced must have strengthened his dedication and enhanced his feeling of responsibility, and this is what kept him driven.

The Montreal Protocol on Substances That Deplete the Ozone Layer and the increasing availability of substitutes for CFCs have combined to change Rowland's situation. He has become a main-stream researcher again, without having to give up what he was doing. When he published his finding about the mercury content in fish, for a while he came to be looked upon as antienvironmentalist. Eventually he was increasingly considered a hero to environmentalists. In reality, during most of his career he considered hazards from chemistry only inside the laboratory, and these were acute hazards, not long-term ones.

In spite of early accusations, Rowland never became a zealous environmentalist; he remained a reluctant one throughout his career. He did not go on crusades; rather, he soberly presented his scientific findings. This was fortunate because he thus appeared more objective than if he had been overly emotional. Harold Johnston was an absolutely trustworthy scientist with impeccable data, yet he appeared too passionate to some when he argued against supersonic transport, and thus his arguments were taken less seriously.[15]

Since 1987, Rowland and Molina's discovery has received the widest possible international recognition. The highlight was the awarding of the Nobel Prize in Chemistry in 1995, which they shared with Paul Crutzen. There were some dissenting voices, however, who considered their Nobel recognition political, but this was relevant only to the extent to which the environment could be considered a political issue.

It was a defining moment when, for the first time, the ozone hole over the Antarctic was detected by hardcore observation and data. Then came the Montreal Protocol, prescribing the phasing out of the production of CFCs responsible for ozone depletion. The Montreal Protocol has been considered one of the most, if not the most, successful international agreements. It was opened for signature in September 1987, and by September 2009, all member states of the United Nations had ratified it. There is hope that by 2050, the ozone layer will fully recover.

CHAPTER 11
MENTAL EXERCISE
Polymerase Chain Reaction

Evolution is and has always been a genetic engineer.

Kary B. Mullis[1]

Kary Mullis (1944–) is sometimes called an antiscience scientist by his peers because of his views and behavior; he often seems to cherish his outcast status. Yet his discovery of the polymerase chain reaction (PCR) in 1983 was above all an intellectual achievement. Because of the technical realization by Mullis and a team of his colleagues at Cetus Corporation, it is possible to reproduce fragments of DNA in an unlimited number of copies and to use them in criminology, medical diagnosis, and in numerous other areas of great import. Mullis might be considered a one-discovery scientist, but while he was a graduate student in biochemistry, he published an intriguing paper in Nature *about fundamental particles traveling ahead and backward in time. This is the same journal that declined to publish his original report of the discovery of PCR—as did the other prestigious journal* Science. *Soon after PCR had become an extraordinarily successful technique, Mullis and PCR started living separate lives that involved disputes, animosities, and patent trials. However, none of this changed the fact that Mullis's Nobel Prize in 1993 was fully deserved.*

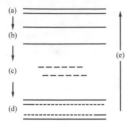

Figure 11.1. Schematic representation of the copying of DNA fragments by the polymerase chain reaction.

(a) Initial stage: double-stranded DNA fragment

(b) The two strands are separated by heating

(c) Synthesized oligonucleotides are added to the container

(d) New two-stranded DNA pieces are formed from the separated strands and the oligonucleotides

(e) Stage "d" has re-created stage "a," but its quantity is doubled; the new DNA pieces are separated by heating (stage "b"), etc.

To his fellow Nobel laureates, Kary Mullis is considered an enfant terrible because of his eccentric behavior, illegal drug use, and antiscience statements; and some even maintain that he should not have received the Nobel Prize. He did, though, in 1993, for his invention of the method of the polymerase chain reaction, or PCR. With even just a fragment of DNA, Mullis suggested a simple but ingenious technique to make an unlimited number of copies from it in a short period of time and with reasonable effort.

Thus, for example, tiny amounts of DNA detected at a crime scene can be multiplied and made suitable for identification. Paternity disputes can be resolved by means of PCR; medical diagnoses have greatly benefited from PCR's possibilities; and PCR can be used in healthcare-related areas, such as inherited disease, infection, and carcinogenesis. Tiny fragments of DNA of long-disappeared fossilized creatures, like dinosaurs, can be amplified and thus made available for further manipulations and the study of evolution. There are virtually unlimited uses of the PCR technique in diagnostics and research. Ted Koppel, the noted television journalist and news anchor, stated about PCR on the *ABC News* program "Nightline": "Take all MVPs from professional baseball, basketball, and football. Throw in a dozen favorite movie stars and a half dozen rock stars for good measure, add all the television anchor people now on the air, and collectively, we have not affected the current good or the future welfare of mankind as much as Kary Mullis."[2]

However, Mullis's reputation among the scientific community has not always been appreciated. Some of his peers have complained about his personality and behavior, and others have even maintained that his contribution was merely a clever technical trick rather than a true intellectual achievement. Even if this were true— which it is not—Alfred Nobel stipulated that the conditions for his prize be that those who receive it should "have made the most important chemical discovery or improvement," and that the contribution was supposed to "have conferred the greatest benefit on mankind."[3] There is nothing in the will about the discovery being intellectual, let alone the discoverer being a great scientist or even a moral person, for that matter.

But Mullis's contribution should be considered a great intellectual accomplishment. He brought together knowledge and information from many different areas of chemistry, biochemistry, biotechnology, and molecular biology, and created something new from these combined fields. One frequent accusation against him has been that all the ingredients he used in his invention were the creations of other scientists. This "accusation" could also be considered eloquent praise for Mullis, because he synthesized those ingredients and produced something that others had not. Even once Mullis's invention came together and he started talking about it to his peers, their reaction to his ideas was lukewarm.

The Mullis family has been in America for generations, though two hundred or so Mullises still live in a village called Flums in eastern Switzerland. When Mullis was awarded the Nobel Prize, the mayor of Flums sent him a congratulatory telegram. Mullis's father traveled as a salesman, selling furniture and also laboratory equipment to high schools. After Mullis's parents separated, his mother went into the real estate business, where she proved herself to be quite successful.

Mullis once said that he did not come from an intellectual background. Living in Columbia, South Carolina, no one noticed that the Mullis family did not have a copy of Erwin Schrödinger's book *What Is Life?* This book, published in 1944, was an important stimulus for physicists and other scientists turning toward the question of heredity. There were not many books in the Mullis home.

The Soviet *Sputnik 1* satellite in 1957 and cosmonaut Yuri Gagarin's orbit around the earth in 1961 played a role in Mullis's intellectual development, if indirectly.[4] The Soviet successes in space exploration gave a big push to American science education and stimulated the interest of American youth in the sciences. At the time of the first Sputnik launch, Mullis was thirteen years old, and he and his friends got engaged in building rockets. He invented a solid fuel from a mixture of sugar and potassium nitrate in a 2:3 ratio by weight and used a dynamite fuse to ignite it. In those days, he was able to buy all the ingredients at a drugstore and hardware store, no questions asked. He put a tiny frog inside a small aluminum canister in the top of his rocket, launched the rocket two miles into the air, and brought the frog back alive.

Mullis graduated in chemistry from Georgia Tech and worked on the nuclear reactor on the campus. He set up experiments that often seemed like something built from an erector set and did routine chemical analyses. In his spare time—and he had a lot of it—he read many years' worth of the popular-science magazine *Scientific American*.

Mullis had a longstanding interest in physics, and he was torn between physics and biochemistry. He thought, however, "that as soon as the Russians and the Americans decided that they were not at war with each other, or one of us was overcoming the other, they won't need as many physicists as they used to." He "thought it was a good idea to go into biochemistry because politicians are always getting heart attacks, and they die of cancer, so they'll need us. But they won't care what's falling out of the sky and they sure don't care how old the universe is."[5]

Mullis was an unusual graduate student in the Biochemistry Department of the University of California at Berkeley. He took astrophysics, but he did not take a course in molecular biology, which he was learning directly from his colleagues. When he read an article by F. R. Stannard in *Nature* about the symmetry of the time axis,[6] his interest was piqued. Mullis prepared an article, which—to his astonishment—*Nature* printed. In it, Mullis suggested that a significant component of the universe was going backward in time. The article was titled "The Cosmological Significance of Time Reversal."[7]

It was a big deal for a graduate student to publish in *Nature*, and as a sole author, no less. It is especially noteworthy that *Nature* published this paper, whereas about fifteen years later it rejected Mullis's account of his discovery of the polymerase chain reaction. Curiously, at both times, the same editor, John Maddox, dealt with Mullis's submissions. Maddox occupied the position of editor of this most prestigious science journal for a total of twenty-two years, during two separate periods, 1966–73 and 1980–95.

When I asked John Maddox about this story, it was obvious that the rejection of Mullis's report remained imprinted on the famous editor's mind. Maddox's problem with the manuscript was that he would have liked Mullis to give an example of application in addition to describing the technique, but Mullis declined without an explanation. In hindsight, Maddox regretted his decision and thought that *Nature* "could do quite a lot to demystify itself, and it ought to be less rigorous in seeking the approval of referees for everything it publishes." Referring in particular to Mullis's case, he said that the journal ought to relax its demands for backing up everything by experiment.[8]

When Mullis completed his graduate work at Berkeley, his committee had to deliberate on whether to let him get away with never having taken any formal course in molecular biology. Finally, he was allowed to pass. This star-studded committee included such members as the Nobel laureate physicist turned molecular biologist Don Glaser and the future editor of *Science* magazine Dan Koshland. Glaser would become one of the founders of Cetus Corporation, where Mullis was to make his seminal discovery. *Science*—under Koshland—would also end up rejecting Mullis's publication on PCR.

As a fresh PhD, Mullis accompanied his wife to Kansas, where she was to attend medical school, and he settled on a job for which he was overqualified. When he returned to California in 1975 with his soon-to-be third wife, he departed yet farther from his field and became the manager of a restaurant and coffee shop. After two years, he secured a job in pharmaceutical chemistry. Toward the end of this period of floating around, between 1972 and 1979, Mullis attended a seminar on DNA, which stimulated his interest in the chemistry of DNA. He started looking for a job that would allow him to be involved in making DNA molecules. He joined Cetus Corporation in 1979, where he was tasked with the synthesis of oligonucleotides. As is well known, DNA consists of billions of nucleotides (units of phosphoric acid, sugar, and nitrogen-containing bases). The term *oligo* refers to linking a few or maybe a few tens of nucleotides together, which is still a far shorter chain than the DNA molecules in living organisms. Eventually, the production of oligonucleotides became mechanized, and machines produced more of these substances than could be used. Mullis started thinking about engaging himself in using rather than making oligonucleotides.

Cetus Corporation was formed in 1971 by a biochemist, a physician (both of whom were MBAs), and a Nobel laureate physicist turned molecular biologist. It was one of the early biotechnology companies in the San Francisco Bay area. A decade would pass before Cetus became a publicly traded company in 1981, but the company was still not making a profit. Still, there was considerable interest in its initial public offering because of the promise of great successes in biotechnology.

It was not long before Mullis became bored with his work and started thinking about various innovations, including the use of computers. In the meantime, there had been developments in the synthesis of DNA, including machines. Mullis was an early user of such machines, for which, again, he kept suggesting improvements. The machines and computerization both led to more free time for

Figure 11.2. Kary Mullis in 2000 in Minneapolis. Photo by and courtesy of Magdolna Hargittai.

Mullis, and he continued to think about simplifications and innovations. His arrival at the concept later to be known as PCR was an outgrowth of this process.

The phosphoric acid and sugar components in DNA are uniform; what varies are the bases. There are four bases belonging to two different classes of molecules: the purines and the pyrimidines (see chapter 1). It is a remarkable property that there are always two forms of any particular sequence of the four bases that can be strung together; the two are each other's complementary forms. If there is one chain, then the other form—the complementary chain—goes backward in sequence. In the double helix, the two chains are held tightly, and to get them apart, DNA has to be heated to a rather high temperature, which would not be practical in life processes. An enzyme can accomplish this separation without heating.

DNA can reproduce itself; this is how it transfers the information of heredity. Its replication could be compared to the structure of layered clay: the next layer of clay is complementary to the previous clay layer, and so on. Any segment of a single helix of DNA has a tremendous affinity for its complementary segment. If there is a short sequence of a DNA segment, say, 20-base-long, in a mixture that has a trillion (10^{12} = 1,000,000,000,000) different pieces, it will find its complementary segment with incredible speed, in less than a minute. The mechanism is not quite clear, but we can imagine it as if the 20-base-long sequence were sliding on all the DNAs in the mixture. The longer the sequence, the longer it takes to find its complementary sequence, but in terms of the human time scale, it is still very, very fast.

The mechanism of finding a complementary pair could be imagined as a 20-base sequence finding not just one complementary counterpart but two, and achieving this in a coordinated manner. Suppose that the short sequence of our example can find its complementary counterpart with a 10^{-6} probability in a pool of 10^{12}

pieces. The number of hits to be expected would be $10^{-6} \times 10^{12} = 10^6$; that is, one million. One million is very few compared with one trillion, but it is still an awfully large number. In order to reduce the number of hits, let us stipulate that the same sequence must be found twice, not just once. In this case, the probability will be much reduced, $10^{-6} \times 10^{-6} = 10^{-12}$, and out of the 10^{12} possibilities, the number of hits in this case would be $10^{-12} \times 10^{12} = 10^0$, which is exactly 1. There will be one single hit.

When using the PCR process for amplification, something will be amplified only if both oligonucleotides find their complementary counterparts. Each of the two oligonucleotides would have a relatively high probability—relative to the total number of possibilities—of making a mistake, but a very low probability of making the same mistake twice. Once the oligonucleotides are identified, they are extended by an enzyme—the polymerase enzyme—that copies the rest of the DNA. The PCR has two functions; that is, two uses. It finds a rare thing, and it amplifies it. The next question is about the mechanism of the process of amplification.

First, the DNA fragment that needs to be multiplied has to be heated; the heat splits the double-stranded DNA into its two strands. Under lower-temperature conditions, the two separate strands each find their complementary partners assisted by the enzyme, DNA polymerase. The process is then repeated, and at each repetition, the number of identical DNA fragments multiplies. Initially, the process was cumbersome due to the thermal instability of the enzyme extracted from the bacterium *Escherichia coli*; this necessitated a new enzyme batch for each new cycle. Eventually, another bacterium, *Thermophilus aquaticus*, was isolated from hot springs; this bacterium was sufficiently stable thermally to allow for continuous repetition of the cycles, and thus the procedure could be mechanized.

Mullis has described the circumstances of how he invented PCR, and however mysterious it sounds, it appears credible. The idea came to him during a night ride in California, but it took months in the laboratory to make it work. One of the occasions on which he talked about the process of the discovery was his Nobel lecture in Stockholm, on December 8, 1993. At the time of his Nobel lecture, he was not employed full time but rather did consulting from his home; hence, for his affiliation, he gave his home address in the printed lecture—something rather unusual for such a publication.[9]

On a Friday evening in May 1983, Mullis and his then girlfriend were driving from Berkeley to Mendocino to spend the weekend at Mullis's cabin. He was thinking about uses of oligonucleotides, and his thinking led him to the basic notions that would come together as the PCR technique. It was a truly intellectual exercise. He wanted to share his idea with his girlfriend, who was versed in chemistry, but he could not wake her up. At a point along Highway 128, he stopped the car because he wanted to make some notes. He continued making notes once they'd arrived at the cabin and over the weekend. Monday morning he was eager to return to his lab to start testing.

Mullis was full of doubts about his new idea. It seemed so simple that he figured others must have thought of it—quite a common doubt among discoverers in similar situations. It was also so simple that he was afraid that when he thought it over, he would find it faulty. Come Monday, he rushed to the library but found nothing that would have shown the same discovery by others. His search was limited, but he must have covered the most relevant literature, because he was searching for DNA and the polymerase enzyme. Then he talked with colleagues during the entire week, and nobody recalled having seen anything similar to his ideas. His colleagues did not pay too much attention, but they may have been accustomed to his wild ideas. Mullis had a whole directory in his computer containing his untested ideas, and now he opened a new file under the name of *polymerase chain reaction*.

Looking back, and knowing how important PCR has become, it is even more surprising that, among all the knowledgeable people with whom Mullis discussed his proposal, nobody took over his ideas. And he was not at all afraid of divulging his ideas; he was not yet sure PCR was as original as he hoped it was. He had the impression that some of his colleagues were bored when he started talking about his PCR in the lab or at parties. The celebrated drive to Mendocino happened in May, and Mullis did his first experiments in September 1983.

On September 9, 1983, late evening, he brought together the experiment that he expected would show the results if PCR worked as he had imagined. On September 12, Mullis went into the laboratory, but he had to be disappointed. Something was wrong, and he realized that he had to deal with many more details than he had hoped for. During the next three months, he did sporadic experiments in addition to his routine duties in the lab.

The first victory occurred on December 16; Mullis had all the details of the experiment worked out and the DNA amplification was successful. He was overwhelmed by excitement. Mullis was alone in the lab and felt the urge to share the news with somebody, anybody— a common reaction in the moment of discovery if it comes at a decisive moment rather than gradually. Mullis ran out into the corridor, and luckily he found someone there who happened to be a patent attorney. Mullis shared the news with him, and the attorney somehow sensed that it might become the most important patent he would ever be involved in at this company. When some unpleasant doubts were later expressed as to whether Mullis alone was *the* discoverer, it came in handy that, at this crucial moment, this patent attorney had witnessed Mullis's joy and excitement.

In the initial stages of the PCR story, however, the problem was not that Mullis's priority was in doubt; rather, it was that hardly anyone would have found his claims for his new concept credible.

Part of this was due to his notoriety in continually coming up with new ideas, and part of it was due to the animosity many of his colleagues felt toward him. Still, Mullis did not keep his idea to himself, and nobody tried to scoop him during the long months between his coming to the idea and the time that Cetus started taking him seriously.

Later, there were claims that Mullis had discovered nothing new, because many of the elements for PCR already existed in the literature. This was indeed true, but these elements had not yet been put together—not even during those long months while Mullis freely talked about his idea. At the patent trial initiated by DuPont against Cetus, the Cetus lawyers asked, "Why had the PCR technique languished, if it had been known for fifteen years?"[10]

It took considerably more effort than Mullis's thought experiments to create the PCR technology, and toward this effort, a team of Cetus associates did the lion's share of the work. They eventually chose a model system for their research, the beta-globin mutation, which causes sickle-cell anemia. This disease was the first so-called molecular disease identified as such by Linus Pauling when he detected a slight difference between the blood protein hemoglobin of healthy and sick people. Furthermore, Pauling pointed to the genetic origin of the difference when he found that it was a familial disease. By the early 1980s, sickle-cell anemia had become well studied; the DNA sequence coding for hemoglobin had been determined.

The difference between the DNA coding for the healthy and sick hemoglobins had been determined to be a single base pair at a known site. A tedious procedure was used in the diagnosis of sickle-cell anemia; this was the kind of work Mullis intensively disliked and which prompted him to seek simpler approaches.

It has already been alluded to that Mullis had difficulties in publishing his discovery. Cetus worked out a complicated scheme to publish PCR, both the concept and the application. The difficulties were compounded by the need to file the necessary patents for the new technology to secure the company's rights. The plan was that Mullis would publish the concept, which he eventually did with his technician as his coauthor. This should have been followed by a multiauthored paper describing the actual application. Because Mullis was slow in writing up his paper, the application, using the beta-globin system as the example, appeared first. On the author list, Mullis was sandwiched between other Cetus associates.[11]

When Mullis was finally ready with his manuscript, he sent it to *Nature*. It was rejected. Unfortunately, Mullis did not explain to *Nature*'s editor that the laboratory realization of his concept had also been worked out but had been submitted to *Science*. Lacking experimental support, *Nature* did not want to publish the article—as we have already seen. Mullis's disappointment was then compounded when *Science* also rejected his manuscript. This rejection was the more puzzling of the two, because the editors of *Science* must have been aware of the successful experimental realization of the concept described in the manuscript. The paper was finally published in a volume of the series *Methods in Enzymology*,[12] a respectable venue, but of far more limited distribution than *Nature* or *Science*.

A great opportunity for Mullis and PCR arose when he gave a presentation at one of the prestigious symposia of the Cold Spring Harbor Laboratory. Another paper, which resulted from this appearance, was published in 1986.[13] By this time, PCR had become a major tool in molecular biology and diagnostics. In the same year, Mullis left Cetus. He complained about the company's curbing his freedom of research, that he had been placed under supervision, and that he had to report regularly to his superiors about the progress he was making and the work he was planning.[14] Cetus had awarded Mullis $10,000 for his invention, which is much more than the usual $1 that authors of patents typically receive from the companies that own the patents. Mullis did not receive anything from the sale of PCR in 1991, neither stock options nor other perks that he might have expected once PCR became a spectacular commercial success.

The controversies over the PCR patent eventually served to Mullis's great benefit. In 1985, a patent for PCR was granted to Mullis, and it was assigned to Cetus. In 1990, DuPont challenged this patent on the basis that H. Gobind Khorana and his associates had published papers in which they had described the amplification of minute amounts of DNA by the enzyme DNA polymerase. Khorana was a Nobel laureate who shared the distinction with two others in the physiology or medicine category in 1968 for their interpretation of the genetic code and its function in protein synthesis. Had Khorana invented PCR—and this would certainly have occurred after the work that earned him his Nobel Prize in 1968— he could have been a viable candidate for a second award. Arthur Kornberg, the Nobel laureate discoverer of the enzyme DNA polymerase, was among the expert witnesses for DuPont.

Mullis and Cetus had now a common goal of showing DuPont's claim untenable, which meant that it was also in the interest of Cetus to show that Mullis was *the* principal discoverer of PCR. The outcome of the trial could be considered, so to speak, a legal seal of approval reaffirming Mullis's role in the discovery. In February 1991, after a torturous legal procedure, Cetus's PCR patents were upheld in court. However, Cetus did not exist for long; in July 1991 it was announced that another pharmaceutical company would buy Cetus. Later in 1991, while Cetus was still independent, the company Hoffmann-La Roche paid about three hundred million dollars for Cetus's PCR technology, roughly half of the total amount for which Cetus would then be sold.

It was as if the Nobel Committee of Chemistry had only been waiting for the outcome of the legal procedure, because in 1993 it awarded Mullis the Nobel Prize in Chemistry. Mullis's former colleagues who helped PCR become the tool it now was had mixed feelings about the award. They were happy to see the first Nobel Prize going for research at a biotechnology company. Also, they surely appreciated the significance of PCR. It must have occurred to them, though, that they had played a significant role in helping PCR to become a practical tool, a great technique, and not merely a brilliant intellectual achievement, which Mullis's contribution certainly was.

The Nobel Prize benefited Mullis a great deal. Never a mainstream scientist, Mullis was brought into the limelight. In his words, "Once you have [the Nobel Prize], there is not a single office in the world that you can't go into. If you call them and say, I would like to talk to you about something, and I'm so-and-so, the Nobel laureate, they'll see me at least once."[15] In the wake of his award, he decided to return to his old hobby of writing and he completed the book *Dancing Naked in the Mind Field*, which was published in 1998. He explained the title of the book by characterizing himself in this way: "I've always been wide-eyed, bushy-tailed, accepting whatever is there until it proved to be no good."[16]

Mullis discusses a broad range of topics and provides a number of anecdotes in his book. Hillary Clinton impressed him with her knowledge of foreign healthcare systems; he called the empress of Japan "sweetie," and he offered the king and queen of Sweden his son as husband to their daughter for one-third of their kingdom. He described his childhood chemistry experiments, his adult experiments with illegal drugs, his "self-healing" experiments, and his nonappearance at the O. J. Simpson trial. Especially unsettling were his comments ridiculing the connections between HIV and AIDS, between fossil fuels and global warming, between CFCs and the hole in the ozone layer, and between diet and health.

Mullis has been engaged in a great variety of occupations before and since receiving his Nobel Prize. Perhaps the side project most intimately related to his discovery was his company Stargene. The company's plan was to sell copies of DNA fragments of famous people, such as Abraham Lincoln, Elvis Presley, Marilyn Monroe, Michael Jackson, and others—anybody whose DNA samples would be possible to acquire. For example, when he and his associates found a hair collection of famous people, they planned to extract DNA out of hair and make copies with PCR. They would build Michael Jordan's DNA into basketball shoes, so people would buy such shoes and jump higher. Mullis compared this gimmick to people buying a pet rock. He had practical difficulties with licensing, though, and he ascribed these difficulties to his inexperience in such matters.[17]

Kary Mullis likes to strike out in many directions in his activities. Lately he has become increasingly interested in how humankind could enhance defense mechanisms from infectious diseases. He explains the clashes of his views with those of mainstream scientists by the fact that he is an outlier. In talking with him, he gives a different impression from the image one gets from what is written about him. He appears gentle and diffident, with a sense of self-deprecating humor. One of his aspirations has been to become a writer. The Nobel Prize paved his way to publish his first and so far only book. I am curious as to whether the lack of a second Nobel Prize means that there will never be a second book. I would not venture, however, to make predictions about Kary Mullis.

CHAPTER 12

DOING WHAT NATURE DID NOT
"Noble" Compounds

When I broke the seal between the PtF$_6$ and xenon, there was an immediate reaction.
Neil Bartlett[1]

The great French chemist Marcellin Berthelot likened chemistry to the arts when he said that chemistry is unique among the sciences because it is capable of creating the subjects of its inquiry. The compounds of noble gases are such a creation. The noble gases themselves were discovered at the end of the nineteenth century, and during the following decades there were several unsuccessful attempts to combine the atoms of the noble-gas elements, such as xenon and krypton, with atoms of other elements, especially fluorine. Neil Bartlett (1932–2008) was the first to succeed doing this in 1962, and his work soon opened up a whole new field with one research group after another reporting the production of new noble-gas compounds with interesting properties and structures.

I n the June 1962 issue of the British journal *Proceedings of the Chemical Society*, Neil Bartlett, then of the University of British Columbia (UBC) in Vancouver, published one of the shortest—less than 250 words—seminal papers in science history. The paper announced the discovery of a compound formed by a

Figure 12.1. Neil Bartlett at the time of his discovery. Courtesy of Christina Bartlett, Orinda, California.

noble gas.[2] Nobody had ever succeeded in preparing such a compound; in fact, the name of these elements—*noble gases* or *inert gases*—referred to their perceived inability to make unions with other elements. The compound—a mustard-yellow solid—was prepared by Bartlett from xenon (Xe) and the highly reactive blood-red hexafluoroplatinate (PtF_6): its formula was $XePtF_6$.

Bartlett first submitted his report to *Nature* on April 2, 1962, but when he did not hear from the journal in a few weeks, he withdrew his report. As he learned later, *Nature* acknowledged the receipt of his submission by surface mail, which didn't arrive in Vancouver until June. By then, Bartlett's manuscript had reached the editorial office of the *Proceedings of the Chemical Society.* He was fully aware of the importance of publishing his discovery as quickly as possible.

The manuscript arrived at the *Proceedings* on May 4. Rumor has it that not everybody found the report about the new substance to be credible, but the editor accepted the paper right away and had it printed in the next issue. The article contains the weighty statement that "this compound is believed to be the first xenon charge-transfer compound which is stable at room temperatures,"[3] but the overall style of the text is low-key. An outsider would not guess that a groundbreaking discovery was being reported. The term *charge-transfer* referred to the nature of bonding between xenon and the rest of the molecule, indicating that the attraction between the positive charge of xenon and the negative charge of the rest of the molecule held it together. One of the consequences of the discovery was that the label *noble gas* has gradually become preferable to *inert gas*. After Bartlett's discovery, the noble gases could no longer be considered inert.

Several ingredients came together for Bartlett's discovery in a fortunate way. These are not mentioned in his published report, in keeping with the usual presentation of research results, which are limited to the dry facts directly related to the successful experiments. The first interesting question in Bartlett's discovery was how the idea of trying to make a noble-gas compound came to him, and how he arrived at the idea of trying the particular reaction between Xe and PtF_6. Bartlett knew about the existence of the compound PtF_6 because of his long-held interest in fluorine chemistry. Fortunately for him, by the time he was about to embark on some of his experiments, Bernard Weinstock and his group at the

Argonne National Laboratory had recently published their preparation of this substance. Bartlett wanted to make his own PtF_6, but he did not find it easy. At this point, he was not yet thinking of involving noble gases in his experiments.

Bartlett was working on the preparation of PtF_6 with his student, Derek Lohman, and their trials led to an accidental preparation of a new substance of the composition of PtO_2F_6; that is, they had inadvertently combined PtF_6 with oxygen. They determined the composition of the new substance in a straightforward way, but it appeared a puzzle to them how the molecule was held together; in other words, how PtF_6 could make the very stable oxygen molecule react with it. It took a lot of experiments, thinking, and convincing doubting colleagues before the team came to the conclusion that the substance could be looked at as being $O_2^+PtF_6^-$; the attraction between the positively charged oxygen part and the negatively charged PtF_6 part kept this substance together.

This meant that PtF_6 had such a strong affinity toward electrons that it was capable of removing an electron from the neutral and very stable oxygen molecule, turning it into an oxygen molecular ion with a positive charge. In the language of chemists, this meant that PtF_6 was a strong oxidizing agent—apparently the most potent oxidizing agent ever discovered. It was serendipitous that Bartlett and Lohman produced O_2PtF_6, but it was no accident that Bartlett understood the nature of its bonding. The realization led him to continue his thinking along this path. He wondered what might be even more difficult to oxidize than molecular oxygen.

It was another fortunate circumstance that, at this time, Bartlett was preparing to teach his undergraduate chemistry course, so he was studying the most fundamental properties of the elements to discuss them in class. Some researchers consider teaching—especially teaching introductory courses—a burden. Others like to go back to the basics and find it a challenge to illuminate the uninitiated in the fundamentals of their science. There may be unexpected benefits from such efforts. In a famous example, Dmitri Mendeleev was preparing his lectures in general chemistry and discovered the periodic variations in the properties of elements. From this, he constructed the periodic table of the elements and gave humankind one of the most beautiful and useful intellectual tools for understanding nature.

As Bartlett was compiling his lectures, he was looking at the energies needed to remove electrons from atoms and molecules, and he noticed that the energy needed for removing an electron from the oxygen molecule was about the same as for removing an electron from a xenon atom. This energy is called the ionization potential because the process converts neutral atoms or molecules into ions. From the similarity of the ionization potentials of molecular oxygen and xenon, it was a short step to the realization that if he succeeded in combining O_2 and PtF_6, it should be possible to do the same for Xe and PtF_6. On paper, this worked fine; now the question was whether he could do the experiment.

When Bartlett realized that the way to go was reacting xenon with PtF_6, he ordered some xenon. His means were limited, however, and he had to restrict his order to 250 cubic centimeters of the noble gas. Although he already had two graduate students, they were inexperienced, so Bartlett had to do even the glassblowing for the apparatus he designed to undertake the experiment. The crucial experiment was performed on March 23, 1962. It was a Friday, and the apparatus was brought together by 7 p.m. Both Bartlett's graduate students had left by this hour, as they lived in a student residence where dinner was served at 6:30 p.m. Bartlett was all alone in the lab.

The moment arrived. Bartlett broke the seal between PtF_6 and Xe, and he did not have to wait long: an immediate reaction occurred. He could see it with his naked eyes, and he was aware of the fact that he was watching something nobody had seen before. Bartlett desperately wanted to tell somebody—anybody—about it.

He went out into the corridor, but there was not a soul in the whole building. So he went back to the lab, where suddenly he was flooded with doubts. "Maybe xenon was impure, maybe there was some oxygen present, maybe I'm just fooling myself."[4] Both the urge to share the news about his experiment and the instant doubts are typical in such moments of discovery.

Bartlett evaporated the reaction product by heating it under vacuum and collected the evolving gas in one part of his apparatus. He then sealed this part of the apparatus, and the next day he gave it to his mass spectrometrist colleague, David Frost. The mass spectrometric analysis is capable of determining the composition of the substance in the gas. Frost's experiments during the next couple of days provided reassuring proof to Bartlett that his product was exactly what he was hoping for.

The world of chemistry appeared to be ready for Bartlett's discovery. Previously, nobody had succeeded in making noble-gas compounds, but now, within a couple of months, a group of scientists at Argonne National Laboratory, Howard Claassen, Henry Selig, and John Malm, reported the synthesis of the first binary xenon compound, XeF_4.[5] The term *binary* referred to the fact that the compound consisted of atoms of only two elements; in this case, xenon and fluorine, whereas Bartlett's $XePtF_6$ was a ternary compound. The successful preparation of further xenon–fluorine compounds was announced before the end of 1962, by Rudolf Hoppe and his group at the University of Münster, Germany,[6] by Jozef Slivnik and his group at the Jozef Stefan Nuclear Institute in Ljubljana (then Yugoslavia, now Slovenia),[7] and by Paul Fields and his associates at the Argonne National Laboratory.[8]

Word was spreading fast about Bartlett's discovery even prior to its publication among those chemists who were interested in such new developments. It is common practice to inform colleagues about new findings. Before the Internet age, printed copies

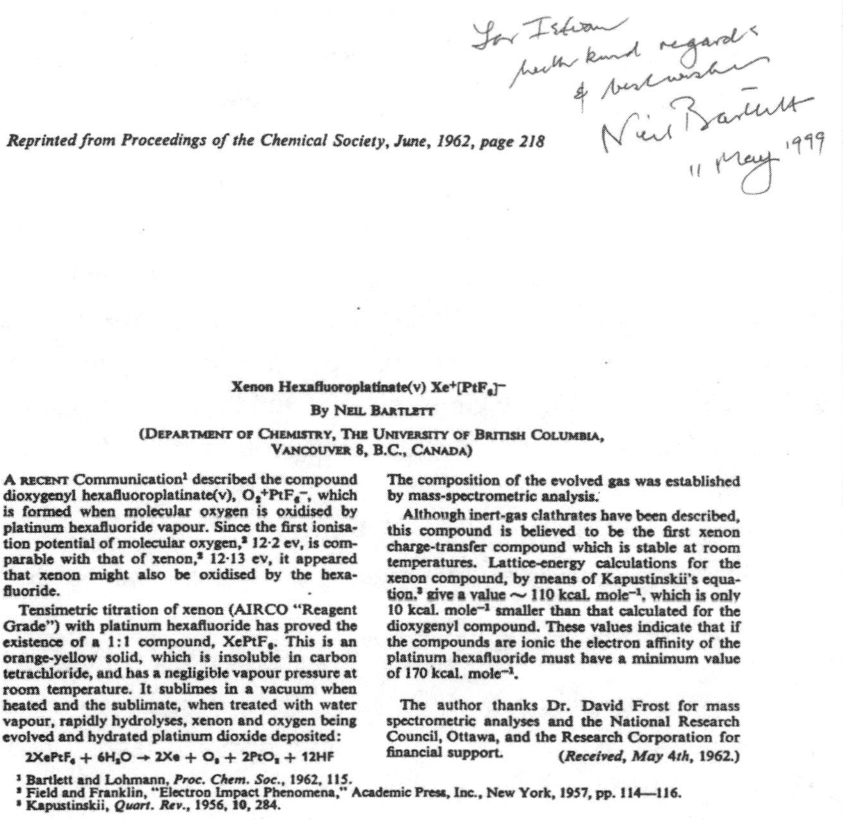

For Istvan
with kind regards
& best wishes
Neil Bartlett
11 May 1999

Reprinted from Proceedings of the Chemical Society, June, 1962, page 218

Xenon Hexafluoroplatinate(v) Xe$^+$[PtF$_6$]$^-$

By Neil Bartlett

(Department of Chemistry, The University of British Columbia,
Vancouver 8, B.C., Canada)

A recent Communication[1] described the compound dioxygenyl hexafluoroplatinate(v), $O_2^+PtF_6^-$, which is formed when molecular oxygen is oxidised by platinum hexafluoride vapour. Since the first ionisation potential of molecular oxygen,[2] 12·2 ev, is comparable with that of xenon,[2] 12·13 ev, it appeared that xenon might also be oxidised by the hexafluoride.

Tensimetric titration of xenon (AIRCO "Reagent Grade") with platinum hexafluoride has proved the existence of a 1:1 compound, XePtF$_6$. This is an orange-yellow solid, which is insoluble in carbon tetrachloride, and has a negligible vapour pressure at room temperature. It sublimes in a vacuum when heated and the sublimate, when treated with water vapour, rapidly hydrolyses, xenon and oxygen being evolved and hydrated platinum dioxide deposited:

$$2XePtF_6 + 6H_2O \rightarrow 2Xe + O_2 + 2PtO_2 + 12HF$$

The composition of the evolved gas was established by mass-spectrometric analysis.

Although inert-gas clathrates have been described, this compound is believed to be the first xenon charge-transfer compound which is stable at room temperatures. Lattice-energy calculations for the xenon compound, by means of Kapustinskii's equation,[3] give a value ~ 110 kcal. mole^{-1}, which is only 10 kcal. mole^{-1} smaller than that calculated for the dioxygenyl compound. These values indicate that if the compounds are ionic the electron affinity of the platinum hexafluoride must have a minimum value of 170 kcal. mole^{-1}.

The author thanks Dr. David Frost for mass spectrometric analyses and the National Research Council, Ottawa, and the Research Corporation for financial support.   (*Received, May 4th, 1962.*)

[1] Bartlett and Lohmann, *Proc. Chem. Soc.*, 1962, 115.
[2] Field and Franklin, "Electron Impact Phenomena," Academic Press, Inc., New York, 1957, pp. 114—116.
[3] Kapustinskii, *Quart. Rev.*, 1956, 10, 284.

Figure 12.2. The first report of the production of a noble-gas compound in June 1962 with a handwritten dedication to the author, 1999. Neil Bartlett, "Xenon Hexafluoroplatinate(V) Xe$^+$[PtF$_6$]$^-$," *Proceedings of the Chemical Society* (June 1962): 218. Reproduced by permission of the Royal Society of Chemistry.

of the manuscript submitted to a journal would be circulated among colleagues; such copies were called preprints. Today, of course, communication of this sort has become much simpler and much faster. At the time of his initial submission to *Nature*, Bartlett informed four people whom he trusted about the manuscript; they were all British scientists, and they realized the importance of Bartlett's discovery. They realized it to such an extent that they found it necessary to inform other colleagues about it, but they talked only with people whom they trusted with holding this confidential information. As a result, the news traveled fast.

The news about Bartlett's discovery encouraged others to go ahead with their own experiments, trying to prepare further noble-gas compounds. Once there was a precedent, it prompted new experiments, because such experiments were no longer considered hopeless. The spreading of the news played a role in a later dispute that developed concerning the priority of the discovery of xenon tetrafluoride, XeF_4, between the German group and the Argonne laboratory. Bartlett referred to this development as follows: "Certainly the Argonne discovery was a direct consequence of mine. I had after all strayed into their territory when I discovered the nature of $O_2^+[PtF_6]^-$. They were all set to repeat my experiment and to extend it to the other transition series hexafluorides. . . . Professor Hoppe [in Germany] has never acknowledged that his efforts were stimulated by my publication of $XePtF_6$, although he has acknowledged my priority."[9] In contrast with other priority disputes in connection with some of the follow-up discoveries, Bartlett's achievement has remained unambiguously recognized as the pioneering one.

In order to truly appreciate Bartlett's achievement, we must remember that he was by far not the first who tried to prepare noble-gas compounds. The earliest attempts go back to the time of the first isolation and identification of the noble gases and to Henri Moissan and William Ramsay at the end of the nineteenth century. Ramsay discovered the noble gases and determined their position in the periodic table of the elements. He worked at that time at University College London and in 1904 he was awarded the Nobel Prize in Chemistry for these achievements. Ramsay's first discovery was that of argon in 1894, and the discovery of fluorine by Moissan in 1886 preceded it by only a few years. In 1892, Moissan invented the furnace with an electric arc, an important apparatus for reaching high-temperature conditions in his chemical experiments. He is considered to be the founder of high-temperature chemistry, which has had great industrial significance.

When Ramsay observed the inertness of argon, he sent a sample to Moissan, who mixed it with fluorine and passed the reaction mixture through an electric discharge. He analyzed the effluent—the resulting product—very carefully and could not detect any sign of any combination. He reported this observation to Ramsay. Due to Moisson's authority, this early experiment had a strong impact on how chemists viewed the noble gases for a long time. Moisson was awarded the Nobel Prize in Chemistry in 1906 for the discovery of fluorine and the invention of his electric furnace. Incidentally, in that year, Mendeleev had been the front-runner for the Nobel Prize, but the final vote was in Moisson's favor. By the time the 1907 Nobel Prize would have been available, Mendeleev had died.

The most noteworthy of later attempts to prepare noble-gas compounds was by a chemistry professor, Don M. Yost (1893–1977), and his graduate student, Albert L. Kaye, in 1933, at the California Institute of Technology. They acted on Linus Pauling's suggestion to try combining xenon and fluorine. Pauling's interest originated from his studies of the nature of the chemical bond; he wanted to have examples of bonding between atoms of the most diverse elements. He was concerned with the ability of the atoms of different elements to attract electrons, a property known as electron affinity. Pauling devised a scale of values, called electronegativites, to characterize the electron affinity of the elements. Fluorine had the largest value; it was so electronegative that Pauling envisioned that it could attract away an electron even from the atoms of the noble gases. It was a hypothesis, though, and Pauling wanted experimental confirmation—hence his request to Yost.

It was not easy to get hold of a xenon sample for the experiment at that time, but Pauling managed to get one from his old chemistry teacher, Fred Allen. Allen had been Pauling's professor in physical chemistry at Oregon Agricultural College. Not only did Allen teach Pauling chemistry; he had him hired as an instructor, which helped Pauling survive hard financial times. By the 1930s, Allen was at Purdue University, where he had a productive career. He also had xenon in his possession.

Yost and Kaye's attempt did not succeed but was found important enough to be described in the *Journal of the American Chemical Society*. This was a rare case when a failure of an experiment found its way into the mainstream scientific literature. Yost and Kaye concluded that "[i]t cannot be said that definite evidence for compound formation was found."[10] Even more significantly, they stated that "[i]t does not follow, of course, that xenon fluoride is incapable of existing."[11]

Bartlett examined Yost and Kaye's work and thought it unfortunate that they used a quartz bulb for carrying out their reaction. When they passed an electric discharge through the bulb, all that likely happened was an attack on the walls of the quartz container. Bartlett supposed that they got silicon tetrafluoride, SiF_4, in addition to oxygen, and that both the silicon and the oxygen must have come from the container material. Bartlett ascribed Yost and Kaye's main mistake to the fact that they did not take their reaction mixture out into the California sunlight, because the strong sunlight might have provided sufficient energy for the reaction to take place.

Three decades later, after Bartlett's discovery of the first noble-gas compounds, Yost looked back and summarized his and Kaye's efforts and the lessons that could be learned from them. He described the rudimentary conditions of their experiments and noted that "only visionary scholars"[12] could have dreamed of the conditions among which decades later similar experiments could be performed. He put it succinctly if symbolically, "Fluorine chemistry was then carried out in the days of wooden ships and iron men."[13]

Yost was magnanimous in evaluating Bartlett's achievement. In some sense, Bartlett defeated Yost and Kaye, and a lesser person than Yost might have belittled Bartlett's discovery. The story of the Greek historian Thucydides comes to mind. He lived four hundred years before the Christian era and became a history writer after the great Athenian statesman Pericles had defeated him in a power struggle. That we know so much about the golden age of Athens is due to Thucydides. He could have written disparagingly about Pericles, out of spitefulness. But he found it more honorable to have been defeated by a great person than by a nobody.

Yost stated that "chemistry had reached a stationary state, during which no profoundly fundamental discoveries were reported or even deemed possible. . . . This state of affairs has now been changed. . . . So long as man shows any interest whatever in chemistry, the discovery of xenon and other noble-gas fluorides will not be forgotten."[14] Further, Yost advanced his vision of the future: "One can envision a whole sombrero full of studies that can be made on noble-gas compounds and their derivatives, which will enrich our knowledge of nature and her laws."[15] He also predicted that applications would follow.

Bartlett considered it a great advantage for his experimentation that he worked with glass and could always follow with his naked eyes what was happening to his reaction mixture. In contrast, the more sophisticated Argonne Laboratory used metal lines and metal valves, so the researchers lacked the ability to observe what was going on during the reaction. Bartlett stressed, though, that he had learned a lot from the Argonne work on platinum hexafluoride, and he praised the beauty and sophistication of their work.

Throughout his career, Bartlett stayed in preparative inorganic chemistry and produced exciting new substances that enriched the field enormously. When asked if there was ever a compound that he would have liked to prepare but never succeeded at, he named gold

Figure 12.3. Neil Bartlett explaining an experiment to Magdolna Hargittai, on his last day at his laboratory in Berkeley, 1999. Photo by the author.

hexafluoride, AuF_6. As he was closing down his laboratory at the Berkeley campus in 1999, he realized that now this task would fall to somebody else, but he gave parting advice: "It should exist, if made at low temperature and kept cold."[16]

Neil Bartlett was a gracious man who was happy seeing other people succeed. He could hardly contain his joy when he learned about the discovery of the new substance consisting of hydrogen, argon, and fluorine, with the formula of HArF, by a group of chemists in Helsinki.[17] He heard about it over the public radio at dinnertime on August 24, 2000. He immediately attempted working out its bonding properties, and the next day he sent an e-mail message to his fluorine chemist colleagues describing his findings. He sent congratulations to the group in Finland and subsequently had a pleasant exchange of letters with them.

The molecular structures of noble-gas compounds have provided beautiful and instructive examples of various models of molecular geometry and chemical bonding. In particular, xenon hexafluoride served as an early proof for the validity and usefulness of a model that has since spread all over the world and is being taught in all high school chemistry and freshman chemistry courses. It is called the valence shell electron pair repulsion (VSEPR) model.[18]

Just by looking at the formula of xenon hexafluoride, XeF_6, one would suppose there to be a symmetrical arrangement of the six fluorine atoms around the xenon in the center. Some rudimentary quantum chemical calculations confirmed this intuitive line of thought. However, the originator of the VSEPR model, Ronald Gillespie, made quite a stir when he warned that his model predicted a less than regular shape for this molecule. Sure enough, experimental studies confirmed what Gillespie's model had predicted. This was at a time when the model was in its infancy, and the story provided a tremendous boost for Gillespie and his model. Today, after many additional studies, we still do not know all details of the exact shape of the XeF_6 molecule.

Toward the end of his career, Bartlett and his group at Berkeley returned to the topic of his original discovery and performed an extensive investigation concerning the nature of $XePtF_6$.[19] It seems that in hindsight he felt that the original investigation was done in some haste, which is understandable, because the discovery must have created euphoric conditions at the time. However, Bartlett liked to get to the bottom of whatever he was involved in, and as a result, he published another paper about this substance—a much longer one than his initial report of the discovery. There were new details about the original reaction, but the essential findings of 1962 remained valid.

It is interesting to look at the paper's list of authors. Two of them were Bartlett's associates at Berkeley. Then there was Narendra K.

Jha, who had been Bartlett's doctoral student at UBC in the early 1960s and who had received his PhD degree there in 1965, but by the time of this more recent paper, he had been long gone from UBC. Nonetheless, he was assigned UBC affiliation, which is especially conspicuous since no other affiliation was given for him. Then, in an apparently romantic gesture, Bartlett added UBC to his own Berkeley affiliation. After his official retirement, in the year 2000, Bartlett gave a full account of his forty years in fluorine chemistry.[20]

*   *   *

At this point, it may be illuminating to share a few words about Neil Bartlett's life and career. He was born in 1932 in Newcastle-upon-Tyne, England. His parents began their married life at the time of the Great Depression, at which time they ran a corner shop. His father had served in World War I and suffered from being severely gassed. He died at a young age, leaving his widow with three children, but she managed to cope. She encouraged her children to study, even tolerating Neil's chemical experiments accompanied by foul smells and frightening bangs. Neil and his brother augmented their pocket money by making and selling ice cream on weekends. Neil used the proceeds to buy chemicals and books.

Bartlett's maternal grandfather had the German-sounding name Vock. He was from Heligoland (at that time a British possession), a small archipelago in the North Sea, some thirty miles off the German coastline. He migrated from Heligoland to Newcastle. When Queen Victoria and the German kaiser exchanged Heligoland and Zanzibar in 1890, Vock became a German subject. Bartlett's mother changed her name from Vock to Voak during World War I to make it sound more British.

Bartlett went to school in Newcastle, England, and then attended King's College, University of Durham, at Newcastle, which is today the University of Newcastle. In 1954, he received his BSc degree, and in 1958 his PhD degree. In 1958, he moved to Vancouver, Canada, became a faculty member of UBC, and stayed there until 1966. It was during this period that he prepared the first noble-gas compound.

Bartlett rose rapidly on the academic ladder at UBC, from lecturer to full professor within five years, but he sensed that his success may have caused some jealousy among his peers. His contact with the Argonne National Laboratory showed him the contrast in opportunities that he lacked in Vancouver. Then, a laboratory accident on January 27, 1963, became forever associated in Bartlett's mind with Vancouver. He and one of his students were looking at an experiment, each wearing a plastic visor, as prescribed by regulations. Although Bartlett had very good eyesight, the visor kept him from getting a good look at the surface of the crystal that was being formed, so he put his visor up, and his student followed suit. At that very moment, the sample they were looking at blew up. Both men were carted off to hospital, where they stayed for over four weeks. Bartlett had some glass in his eye that continued to bother him for the next twenty-seven years; it finally came out in 1990. When he told me about the accident, he stressed how lucky it was that his student suffered much lesser injuries than he did.

There were other unfavorable conditions at UBC at the time, so when an attractive offer came from the prestigious Princeton University, Bartlett accepted. Eventually he realized that he felt more comfortable at UBC than at Princeton, that he preferred a smaller campus to a big one, and that he was a West Coast person rather than an East Coast one. When he received an invitation from Berkeley, he accepted it and moved to California in 1969. There he was professor of chemistry at the University of California at Berkeley and also served as principal investigator at the Lawrence Berkeley National Laboratory. When my wife and I visited him on

May 11, 1999, he did not strike us as a happy person. The date was significant, because it was the day he was closing down his laboratory for good.

Neil Bartlett was an internationally renowned scientist. Among many other distinctions, he was elected fellow of the Royal Society (London) in 1973; he was a foreign associate of the US National Academy of Sciences and of the French Academy of Sciences. In 1976, he received the Robert A. Welch Award in Chemistry. In 2006, the American Chemical Society and the Canadian Society for Chemistry honored him with a wall plaque that identified as an International Historical Chemical Landmark the building at UBC where his seminal discovery took place in 1962. It reads, both in English and in French,

NEIL BARTLETT AND REACTIVE NOBLE GASES

In this building in 1962 Neil Bartlett demonstrated the first reaction of a noble gas. The noble gas family of elements—helium, neon, argon, krypton, xenon, and radon—had previously been regarded as inert. By combining xenon with a platinum fluoride, Bartlett created the first noble gas compound. This reaction began the field of noble gas chemistry, which became fundamental to the scientific understanding of the chemical bond. Noble gas compounds have helped create anti-tumor agents and have been used in lasers.

In 2003, the magazine of the American Chemical Society, *Chemical and Engineering News*, conducted an international survey to select "the 10 most beautiful—that is, elegantly simple but significant—chemistry experiments."[21] Bartlett's experiment in which he produced the first noble-gas compound was among the top ten and turned out to be one of the few among the top ten that unquestionably satisfied the original criteria.

Neil Bartlett figures, although not by name, in the first page of the first chapter in the book *The Periodic Table* by Primo Levi—the Italian chemist turned respected writer: "As late as 1962 a diligent chemist after long and ingenious efforts succeeded in forcing the Alien (xenon) to combine fleetingly with extremely avid and lively fluorine, and the feat seemed so extraordinary that he was given a Nobel Prize."[22] Bartlett never received the Nobel Prize, but this distinction from Primo Levi was yet another indication that his discovery was immortalized in the annals of science.

Figure 13.1. Young Leo Szilard (right) in the company of two future Norwegian Nobel laureates, Odd Hassel (left) and Lars Onsager (center) in Berlin, 1924. Photo by Johan P. Holtsmark, courtesy of the late Otto Bastiansen.

CHAPTER 13

TO SAVE HUMANKIND
Nuclear Chain Reaction

The most important step in getting a job done . . . is the recognition of the problem.
Leo Szilard[1]

Leo Szilard (1898–1964) wanted to use science to protect humankind. In 1939, his letter to President Franklin D. Roosevelt (signed by Albert Einstein) eventually initiated the Manhattan Project. He was different from many of his peers. Szilard considered science to be secondary to world affairs. Yet even in his nonchalant approach to science, he managed to make important discoveries and helped numerous others to achieve success. In the early 1920s, Szilard wrote a paper considered to be the earliest contribution to the field of information theory. In 1933, he discovered two concepts that pointed the way toward harnessing nuclear energy and building atomic bombs. One was the nuclear chain reaction, and the other was the critical mass. He did not follow up these discoveries with experimental work, but when the discovery of nuclear fission by others opened the way to the realization of Szilard's concepts, he performed a crucial experiment confirming that the number of emerging neutrons is larger than the number of incoming neutrons; hence the nuclear chain reaction is feasible.

Leo Szilard came from an upper-middle-class Jewish family in Budapest. He was a creative child with an inquisitive mind and an addiction to truth even when it was inconvenient, but he was not very clever with his hands. He attended one of the prestigious high schools in Budapest and performed superbly in the physics and mathematics competitions for high school graduates. He began his university studies in engineering at the Budapest Technical University, but World War I, the ensuing revolutions, and the vicious anti-Semitism in Hungary interfered, and at the end of 1919, he immigrated to Germany.

In the 1920s, Germany was a flourishing democracy, and Berlin was one of the capitals of modern physics. Szilard found his natural habitat at the physics colloquia of the University of Berlin, organized by Max von Laue, where Albert Einstein, Max Planck, and other luminaries of science were making a collective effort to keep abreast of the recent developments in their field. Szilard was not intimidated by the presence of the greats and communicated with them about everything he was interested in. He told Planck, one of the most creative theoreticians of the century, that he only wanted to get the facts and that he would create the theories himself. Szilard became friendly with Einstein, and they jointly filed a series of patents describing a refrigerator based on new principles.

While Szilard was working on his doctoral dissertation, he recognized an application of the second law of thermodynamics that was broader than had ever been considered possible. He felt it prudent to show his thesis to Einstein before he submitted it. Einstein quickly overcame his initial disbelief and approved of Szilard's conclusions. Szilard's dissertation and his follow-up work included the concept of information transfer, though not by name, and three decades later it was realized that his concept was the earliest contribution to the emerging field of information theory.[2]

Szilard had many other original ideas, but in most cases, he did not follow them up. Though he went on record with some of these ideas, he preferred filing patents to publishing research papers. He was also very good at selflessly giving advice to others and pushing his fellow scientists in the right direction. Some may have resented that his advice often sounded more like instructions, but those who had known him got used to his style, which was characteristic of him throughout his entire life.

Szilard had a keen sense of politics and world affairs, and he had the rare ability of foresight and the even rarer ability to act upon his foresight. Even as a young man, he predicted that World War I would end with the defeat of the Central Powers (Germany and Austria-Hungary) and Russia, which was the more remarkable because the two were on opposite sides of the hostilities. He also foresaw the deterioration of the situation in Germany at the time of the Weimar Republic.

When the Nazis came to power, Szilard devoted a tremendous amount of effort to rescuing scientists and finding jobs for them, regardless of the fact that he was in the same perilous situation that they were. On the British side, the great physicist and discoverer of the nuclear atom model Ernest Rutherford was taking "the lead in opening English academic life to Jewish refugees."[3] It was at this time that Szilard formulated his priorities; he gave preference to nonscientific matters, such as helping other scientists and the defense of democracy, rather than to science and to securing his financial independence.

Szilard mostly stayed in London after fleeing Germany in 1933 and commuted between the United States and England. He vowed that he would stay in Britain until one year *before* the beginning of the war in Europe. This was another example of Szilard's foresight: as soon as he heard about the Munich Agreement between Nazi Germany and Great Britain and France, signed on September 30,

1938, and praised by the British prime minister as an assurance of peace, Szilard thought otherwise and decided to stay in the United States. World War II broke out on September 1, 1939.

It was during his stay in London that Szilard discovered the concept of nuclear chain reaction. One day in 1933, he was walking along Southampton Row when a traffic light made him stop. At that

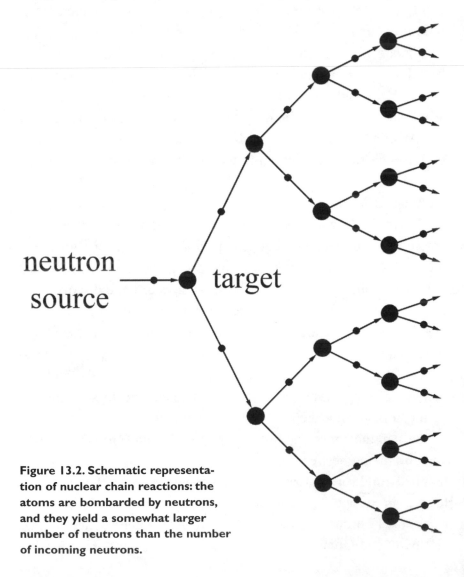

Figure 13.2. Schematic representation of nuclear chain reactions: the atoms are bombarded by neutrons, and they yield a somewhat larger number of neutrons than the number of incoming neutrons.

moment the idea came to him that "if we could find an element which is split by neutrons and which would emit *two* neutrons when it absorbed *one* neutron, such an element, if assembled in sufficiently large mass, could sustain a nuclear chain reaction" (italics in the original).[4]

This idea, in fact, consisted of two concepts, one of which was the nuclear chain reaction. The concept of a chain reaction had been known in chemistry, although there is no evidence as to whether Szilard had been familiar with it. Chain reactions in chemistry were discovered in the 1910s in Germany, and the theories of chemical chain reactions and combustions were worked out in the 1920s and 1930s in Russia and England. Two of the principal scientists in these efforts were awarded the Nobel Prize in Chemistry in 1956. It was Szilard's discovery that nuclear processes could also become chain reactions.

The other concept Szilard advanced was that of critical mass, the smallest mass of the material that could lead to an explosion in a nuclear chain reaction. The chain reaction could be sustained in an amount less than the critical mass of the substance, and in this case the harnessing of energy from the nuclear chain reaction would be possible. If there is a critical mass or more of this substance, then the nuclear chain reaction becomes self-sustaining and ends with an explosion, which points to the possibility of military use; that is, the atomic bomb.

When Szilard came upon the ideas of nuclear chain reaction and critical mass, there still remained unanswered questions about the practical realization of either energy production or producing a bomb. The principal unanswered question involved the identification of the element or elements that would be capable of sustaining a nuclear chain reaction in the first place. But even without such information, the discovery was a milestone, and Szilard considered it sufficiently important to file a patent for it in 1934, assigning it to the British Admiralty somewhat later. This step made it especially obvious that he could not have expected material gains from the patent at that time—and that this was not his purpose.

The atomic age began about three-quarters of a century ago, and the existence of the nuclear chain reaction has become common knowledge in that time. However, when the concept was pronounced for the first time, it was not only shocking, but many experts absolutely refused to consider it as a possibility. Coincidentally, at about the time of Szilard's realization of the concept of a nuclear chain reaction, Rutherford gave a lecture in London on September 11, 1933. Rutherford issued this warning: "[T]o those who look for sources of power in atomic transmutations—such expectations are the merest moonshine."[5]

Szilard was puzzled by Rutherford's statement because he did not think anyone could know what someone else might invent. On June 4, 1934, Szilard visited Rutherford, but when he told the great scientist about his idea of a nuclear chain reaction and that he had already patented the concept, Rutherford became very upset and threw Szilard out of his office. Rutherford's intuitive foresight was legendary, but on this occasion he was wrong—very wrong. In fact, this misjudgment was so much in contrast with Rutherford's usual intuition that it has become popular to attribute his warning to something else, as if he had been trying to divert attention from practical applications of nuclear physics fearing its military use. But such an approach would have been utterly alien to his personality.[6]

Szilard was undeterred. He knew that further experimentation was necessary, and for some of it he enlisted nuclear physicist Lise Meitner in Berlin. This decision is surprising in hindsight, because eventually Szilard would be the most vocal proponent of secrecy in nuclear matters lest Nazi Germany learn from the Allies' efforts. In this case, it would hardly have been possible for Meitner to keep her work secret from German officials, whatever findings she derived. But Szilard kept the main thrust of the concept of the nuclear chain reaction in the patent at the British Admiralty.

For years, the principal missing piece of information continued to be the identity of the element or elements that would be suitable to use for a nuclear chain reaction. Szilard had guesses, including indium and uranium, but he did not know for sure. He might have systematically carried out a long series of painstaking experiments, taking every element in the periodic system to task, but he lacked the means to do this, even if he had been inclined to perform such tests. At the same time, in 1934, one of the greatest physicists of the twentieth century, Enrico Fermi, was conducting experiments at the University of Rome that were highly relevant to Szilard's nuclear chain reaction, although nobody realized this yet. Fermi and his group tested one element after the other—whatever they could get hold of—and bombarded each element with neutrons. When uranium was used as the target, Fermi could have come up with a seminal discovery, but he did not.

Fermi and his associates thought they had produced two new transuranium elements. The fascist media made much of the ostensibly great discoveries, and the two would-be new elements were given ancient Italian names. What really happened was understood only years later. Fermi had split the uranium atoms into two elements of similar masses, but he missed this discovery. It fell to Otto Hahn and Fritz Strassmann to carry out a similar experiment five years later, in December 1938. Lise Meitner and Otto Frisch, who by then were refugees from Nazi Germany in Scandinavia, interpreted the experiment as nuclear fission in January 1939, and quickly published their findings. Soon there were indications that for every one neutron, there were more than one emitted in this nuclear reaction. If they could be sure of this, it would mean that Szilard had the element for the nuclear chain reaction.

It is painful to think what might have happened had Szilard carried out the systematic search for the element suitable for a nuclear chain reaction and found it in 1934, or had Fermi understood the reaction when he bombarded uranium with neutrons in 1934. It is quite probable that Nazi Germany might have built the first atomic bomb and that Hitler would have blackmailed the democracies with it. This is why, after World War II, Szilard suggested, half-joking, that Fermi and he should have received the Nobel Peace Prize, not for something they had discovered but for something they had both missed in 1934.

By the time the question of the number of neutrons was to be answered, both Fermi and Szilard were at Columbia University in New York. Fermi arrived with a fresh Nobel Prize and took up a prestigious professorial appointment there. Szilard was merely a visiting scientist who was allowed to conduct some research at the Physics Department. There was no support for his experiments, so Szilard had to borrow $2,000 from a friend. This was in the spring of 1939, and later in the year, Szilard had to ask his lender, Benjamin Liebowitz, to write off the loan as a bad debt. Eventually, however, Szilard repaid the debt to Liebowitz.

Szilard's experiment was to confirm some French reports from Paris that statistically there was more than one neutron being emitted following the fission of uranium under the impact of one neutron. This was one of the rare occasions when Szilard not only designed the experiments but carried them out, uncharacteristically, with his own hands—and with practical help from Walter Zinn. Szilard could soon report to his friends that he had found the neutrons.

The nuclear chain reaction was no longer a subject for daydreaming, but there were still many practical steps to be taken to bring the concept to realization. Szilard and his team had to be sure that, during the experimentation, the reaction would not end in an unwanted explosion. They needed a suitable moderator that could be inserted within the uranium mass to regulate the reaction. Szilard got the idea that it should be pure graphite. His concern about the purity of graphite was justified. The failure of the German experiments was eventually attributed in great part to impurities in the graphite they had used. At the time, however, ominous signs were arriving from Germany that the importance and potentials of the nuclear chain reaction had been recognized there, too, and that the Germans might be acting on this recognition.

When the summer of 1939 arrived and there was still no funding at Columbia University for the continuation of the experiments, Szilard found the situation increasingly frustrating. The Germans could overrun Belgium at any minute, and by doing so, they would take possession of the Belgian Congo with its vast uranium resources. Szilard and his fellow Hungarian physicists, first Eugene Wigner and then Edward Teller, wanted to make the American administration become aware of the looming danger of the possibility of Hitler's acquiring a new weapon of heretofore unprecedented power. Finally, Szilard proposed and drafted a letter to President Franklin D. Roosevelt, and Albert Einstein signed it. Einstein had not thought that nuclear power could be harnessed in his time, but Szilard convinced him that it could happen.

In their approach to the looming danger that the Nazis might acquire an atomic bomb, both Szilard and Fermi were conservatives, but there was a big difference between Fermi's conservatism and Szilard's.[7] In a discussion sometime in 1939, prior to the realization that uranium would be suitable for nuclear chain reaction, Fermi characterized it as a remote possibility, by which he meant it had a 10 percent probability. To Szilard, however, 10 percent probability of the possibility of a lethal weapon meant a great danger; this was Szilard's conservatism. In his words, "Fermi thought that the conservative thing was to play down the possibility that this may happen, and I thought the conservative thing was to assume that it would happen and take all the necessary precautions."[8] For Fermi, science was his life; for Szilard, science was for the sake of life.

World War II broke out on September 1, 1939, and the United States remained neutral until it was attacked by Japan at Pearl Harbor on December 7, 1941. Everything accelerated once the United States became a combatant on December 8, 1941. The coordinated effort to create an atomic bomb, called the Manhattan Project, began in September 1942. The first experimental nuclear reactor, called a *pile*, was put into operation on December 2, 1942, in Chicago. Szilard played a pivotal role in this, having codesigned the reactor with Fermi, but of equal importance was his involvement in mobilizing his fellow scientists to join the efforts for the atomic bomb project.

Szilard was active in setting up security measures for their program, and he helped convince the scientists to refrain from publishing their results in nuclear physics. The security measures, however, almost barred Szilard, Fermi, and other "alien" physicists from participating in the work. The head of the Uranium Committee, the first organization initiated by President Roosevelt to deal with the atomic project, did not want to include these scientists, because, as he explained, the subject of discussion was secret. The irony was that it was Fermi and Szilard who had invented the process that now had to be kept secret. Gradually, however, the atomic bomb project began to take shape. In 1943, the stage was set for the first atomic bombs to be designed and built at Los Alamos. By then, Szilard, who was operating in Chicago, was no longer involved in the main thrust of the efforts.

It follows from Szilard's personality that he changed his mind about the necessity of deploying the atomic bombs during World War II. As soon as the organization for creating the atomic bomb within the framework of the Manhattan Project came together, Szilard started thinking about the postwar role of nuclear weapons. He anticipated a nuclear arms race among the winners of the war. As early as 1942, he drafted a memorandum about "winning the peace" after the war with Germany. He considered international control to prevent proliferation of nuclear weapons and even preemptive wars to enforce such prevention. These considerations seem as timely at the dawn of the twenty-first century as they seemed to him in 1942. Furthermore, he anticipated the Cold War that would come after World War II, calling it an "armed peace."[9]

Szilard was in constant fear during the initial period of the war that the Germans might be ahead in the race for the atomic bomb. By the time the actual deployment of the first atomic bombs became a possibility, Germany had been defeated, and the threat of a German bomb no longer loomed over the Manhattan Project. From this point on, Szilard no longer wanted to see the bombs dropped. He warned that using the bombs would strengthen the Soviet Union's efforts for an arms race. In this, he proved to be mistaken, as the Soviet Union was preparing for its nuclear weapons program regardless of the intentions of the United States.

Furthermore, Szilard warned of the possibility of the proliferation of atomic bombs and the possibility of small atomic bombs being smuggled into the United States for the purpose of destroying American cities. This proved to be a prescient warning, even if no such event has yet taken place; its eventuality still cannot be excluded. Another of Szilard's prescient visions involved long-range rockets with nuclear warheads attacking American cities. Szilard's suggestion to prevent this was international control, though he was enough of a realist to recognize that even the international control would be difficult to establish if the world had not seen what the atomic bombs could do. Referring to the anticipated postwar situation, he noted that "it will hardly be possible to get political action . . . unless high efficiency atomic bombs have actually been used in this war and the fact that their destructive power has deeply penetrated the mind of the public."[10]

Szilard opposed the use of the atomic bomb from a moral point of view. When he understood that he could not prevent the dropping of atomic bombs over Japan, he and a large group of his colleagues at least wanted to go on record with their opposition to such use. This was a remarkable moment in the history of the atomic bomb: The person who had initiated its development was now recording his opposition to its deployment. When Szilard realized that the atomic bomb could save a large number of American lives, he modified his stand and suggested giving the Japanese suitable warning and an opportunity to surrender. We know that this did not happen, and it is almost impossible to imagine a different scenario from what actually occurred.

After World War II, Szilard enjoyed his celebrity status as one of the pioneers of the nuclear age and helped to create legislation for the civilian control of nuclear matters in the United States. He remained at the University of Chicago in a unique appointment of being a half-time professor of biophysics and a half-time adviser to the innovative president of the University of Chicago, Robert M. Hutchins. Szilard's interests were wide-ranging; he never settled on a topic for long; he never set up a household; and even after his marriage, he never stayed with his devoted wife for very long. French biologist François Jacob thought that for Szilard, the most fitting job would have been "a special job of a bumblebee in a communication system whose task would be talking with people and getting and disseminating news."[11] French biochemist Jacques Monod acknowledged the benefits from Szilard's advice in his Nobel Prize–winning research when he referred to Szilard's "penetrating intuition" in his Nobel lecture.[12]

Szilard had already become interested in biology before the war, and after the war he took up molecular biology for a while. In this endeavor, he was part of a trend of well-established physicists who turned to biology, seeing it as a new frontier where fundamental questions had to be solved and where there was a notion that the solutions would come from outside of biology. Szilard's broad education and outlook helped him to establish himself quickly at the forefront of some biological problems, and he cross-fertilized ideas and interacted with several of the leading biologists of his time. He established a laboratory, attracted an enthusiastic and knowledgeable young associate, Aaron Novick, and together they made innovations in creating a continuous source of bacterial population for research purposes and discoveries in the areas of regulation of gene expression. After a few years, however, Szilard became restless again, and he returned to the main arena of his postwar activities: the politics of arms control and possibilities of peaceful coexistence with the Soviet Union.

Throughout the 1950s and early 1960s, Szilard proposed ideas

Figure 13.3. Leo Szilard at a press conference on nuclear energy at Oak Ridge National Laboratory. From left to right, Walter H. Zinn, Leo Szilard, Eugene P. Wigner, Alvin M. Weinberg. Courtesy of Oak Ridge National Laboratory, managed for the US Department of Energy by UT-Battelle, LLC.

in the arenas of both politics and science that were often viewed with disbelief initially. Many of these ideas, however, were implemented later in his life or after his death. His efforts and those of others contributed to the creation of the Salk Institute for Biological Studies, the Pugwash Conferences on Science and World Affairs, the National Science Foundation, and the European Molecular Biology Organization and European Molecular Biology Laboratory. He was involved in issues that gained recognition only later, such as overpopulation, poverty, the protection of the environment, alternative food supplies, and the dangers of cholesterol.

Szilard sought ways to help bring together the United States and the Soviet Union. He wrote letters to Joseph Stalin and corresponded with Nikita Khrushchev. Szilard had personal discussions with the Soviet leader and even gained Khrushchev's assent to establish a Moscow-Washington "hot line" to avert nuclear catastrophe. But he had no illusions about the nature of the Soviet Union, nor did he doubt that he could not have survived under Soviet conditions. At the time of the debate as to whether the United States should develop the hydrogen bomb, Szilard rooted for Edward Teller's success in convincing the American administration that the bomb was unavoidable.[13] He supported the efforts for developing a "clean" hydrogen bomb that would minimize radioactive fallout. But he was frightened by the new weapons of ever-increasing power with such potential for devastation.

He hoped that eventually the terrifying weapons would scare humankind from considering using them. In a nationally broadcast radio discussion, he proposed adding cobalt to the hydrogen bomb, which would make it yet more dreadful by generating strong and long-lasting radioactivity that would annihilate the human race.[14] This was one of his ways of telling people to stop the arms race. Other notable personalities, such as Alfred Nobel and Linus Pauling, who were also dedicated to eliminating weapons of mass destruction, made similar proposals of creating the most terrible weapons to prevent their use. Albert Einstein recognized early enough that "atomic energy . . . may intimidate the human race into bringing order into its international affairs, which, without the pressure of fear, it would not do."[15] The policy of mutually assured destruction, or MAD, was an embodiment of such a concept, even if it was based on overkill by stockpiling an unprecedented amount of lethal weapons that could have blown up the earth many times over. It can be said, in hindsight, that the policy of MAD helped maintain the peace between the two superpowers for decades. Szilard did not advocate the complete elimination of nuclear weapons, only a drastic reduction and an agreement excluding their use.

High on Szilard's list of concerns was providing the US administrations with the proper advice and counsel from scientists. He'd had this concern since the possibility of the atomic bomb, and this establishment of channels for scientific advice was among the purposes of Einstein's 1939 letter to President Roosevelt. President Dwight Eisenhower recognized the importance of scientific advice, but his institutionalized committees did not involve Szilard. Szilard's role was self-defined, and he wondered whether there was a "market" for his wisdom in the Kennedy administration. Apparently there was, although he did not have direct contact with the president, nor did he have any official capacity for voicing his opinion and proposals.

Senator John F. Kennedy had recognized Szilard's services to the United States and addressed him in a letter of May 27, 1960: "This country owes many debts to you, not only for your scientific achievements but for the great responsibility and imagination you have brought to the problem of securing peace."[16] Kennedy sent his letter to Szilard in hospital, where he was directing his own treatment against bladder cancer. Szilard had declined surgery and opted instead for a drastic radiation therapy, which he and his physician wife had determined to be appropriate. The treatment led to the complete elimination of his cancer. When Szilard died in 1964 from a massive heart attack, the autopsy found no trace of any cancerous growth where his bladder had been.

Szilard lived in two worlds throughout his adult life. He scrutinized every development in science from the point of view of the possible consequences of those developments. In one of his worlds, he tried to predict what was going to happen concerning not only science but everything else as well. In his other world, he fought for what he hoped would happen. This second world took precedence over the first one and became the source of his activism. When the idea of the nuclear chain reaction came to him, Szilard at once felt the tremendous responsibility of being in the possession of a concept that might lead to the most devastating calamity in the wrong hands, but that might also become instrumental in saving the democracies in their fight against evil.

CHAPTER 14

CREATING UNDER PRESSURE
Thermonuclear Explosion

He took pride in having invented the hydrogen bomb and it was easy to declare him a public enemy.
Freeman Dyson[1]

Edward Teller (1908–2003) was not the type of person who would freeze under conditions of great tension. On the contrary, his mental capacity seemed to be enhanced by stress. During the quest for the solution to the thermonuclear bomb, or hydrogen bomb, there was a period of increasing frustration because for years the solution seemed to elude the scientists at Los Alamos. As Teller was the central figure among the proponents of the hydrogen bomb, all attention was focused on him, and, consequently, all blame would be directed toward him for a sustained failure. Under these conditions, whether entirely on his own or sparked by a suggestion made by his colleague Stanislaw Ulam, Teller advanced an idea that solved the debacle. Nuclear physicist Hans Bethe would call this solution as important as the discovery of nuclear fission.

I t seldom happens that a scientist looking for a particular discovery actually finds it. Problems can be solved, but real discoveries often come about unexpectedly. This is why important

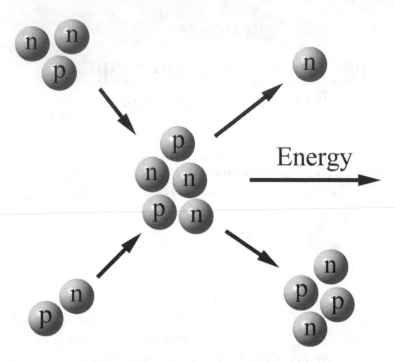

Figure 14.1. Schematic representation of a thermonuclear—or fusion—reaction: in this example, a deuterium nucleus (one proton plus one neutron) and a tritium nucleus (one proton plus two neutrons) are fused, yielding a helium nucleus (two protons plus two neutrons), a neutron, and energy.

discoveries are sometimes overlooked, and only in hindsight does it become apparent that they had been missed. It takes a strong individual to be able to concentrate on a task and find the solution under tremendous pressure. We see this kind of thing happening in the movies: A complicated puzzle needs to be solved in order to save a plane, a ship, a town, or humankind, and the hero solves the problem within a second or two before the fatal explosion. Real life is quite different. However, Edward Teller's discovery of the solution for the thermonuclear bomb was an exception. The tension was high, yet he devised a solution that Hans Bethe compared to the discovery of nuclear fission in terms of its significance.

Edward Teller was one of the most controversial scientists of all time, and the controversial views about him have hardly decreased in intensity since his death. But even Teller's most dedicated enemies never doubted that he was an outstanding physicist. He started his studies in Budapest, then moved to Germany, where he did his doctoral work under Werner Heisenberg in Leipzig in the late 1920s. He worked with Max Born, James Franck, and others in Göttingen in the early 1930s. After he was forced out of Germany in 1933, he spent some time in Niels Bohr's Institute in Copenhagen and at University College London. He continued his interactions with leading physicists, such as Lev Landau and George Gamow. In 1935, he moved to the United States and served as physics professor at George Washington University, during which time he published important papers in molecular physics, atomic physics, and, increasingly, nuclear physics.

He collaborated with others, and his oeuvre thus shows a remarkable diversity of topics and coauthors. The mere enumeration of some of his research achievements is overwhelming, and it includes the first estimation of the substantial energy barrier hindering the rotation around the carbon–carbon bond in the ethane molecule; the establishment of a widely applicable expression to describe multilayer adsorption of gases on solid surfaces; the changes of symmetry in molecules with an incomplete set of electrons; the resolution of a controversy in interpreting the benzene structure; the perfection of the interpretation of beta decay in atomic nuclei; the description of nuclear reactions in the stars; and the discussion of cosmological phenomena; among others.

At the dawn of World War II, Teller became involved in the preparation of the first atomic bombs. He played an active role at the final stage of this work within the framework of the Manhattan Project at Los Alamos. He contributed to solving the problem of implosion, which was used for one of the two principal designs of the atomic bomb. During the war years, Teller's attention was preferen-

tially directed toward the possibility of using thermonuclear reactions in developing a yet more powerful weapon than the atomic
bomb. The atomic bomb was based on nuclear fission in which the
heavy uranium or plutonium is bombarded with neutrons; nuclear
fission produces two other atoms of comparable weight, statistically
more than one neutron, and a large amount of energy.

In contrast, the thermonuclear reaction is the fusion of two
atoms of the heavier isotopes of hydrogen, and in the process, a
large amount of energy is liberated. Thus, one deuterium (D) and
one tritium (T) fuse together, yielding one helium (He), a neutron
(n), and energy:

$$D + T \rightarrow He + n + energy.$$

The name *thermonuclear* refers to the very high-temperature conditions needed for such a reaction to happen. Whereas the principles
of making the atomic bomb were known by the time the Manhattan
Project began, there was no feasible approach yet to producing the
fusion bomb, called also the Super, or the hydrogen bomb. The work
on creating the hydrogen bomb intensified at Los Alamos after the
successful deployment of the atomic bombs over Japan but came to
a halt at the end of World War II. Teller was among those, however,
who thought that no time should be wasted in the development of
the hydrogen bomb lest the Soviet Union blackmail the Free World
with its nuclear weaponry. The proponents of the hydrogen bomb
were sure the Soviets were developing it.

However, Teller was in a minority among the scientists, as most
thought the atomic bombs for which the United States held a
monopoly for the time were sufficient to deter any other threat. Furthermore, it was thought that the Soviet Union, being a rather backward country, would not be capable of developing nuclear weapons
for quite some time. Teller did not underestimate the profession-

alism of the Soviet physicists and knew that a totalitarian regime could focus its efforts in the direction of selected well-defined tasks. Teller at this time was not a member of any influential advisory board or similar bodies, so he launched essentially private lobbying efforts to convince the military and political leaders of the United States that it was necessary to develop the hydrogen bomb.

The General Advisory Committee (GAC) of the recently organized Atomic Energy Commission (AEC) was headed by the father of the atomic bomb, J. Robert Oppenheimer, and counted among its members several leading atomic scientists. This committee took a stand in no uncertain terms against the development of the hydrogen bomb. The members were united except for slight nuances of difference in their opinions. Some suggested that the United States should show an example of restraint and thus force Stalin's empire to follow suit. Others advocated making a solemn pledge not to proceed in the development of the hydrogen bomb. The GAC finally advised the AEC against the development of the thermonuclear weapon, and the AEC thus advised the president. President Harry Truman's leading military and political advisers, however, were for building the hydrogen bomb, and Teller and a few other scientists helped them to acquire the necessary information. Teller was certainly the most visible and most dedicated among them.

It was not known at the time, but the Soviets were already working full steam ahead on their nuclear weaponry. They prepared a report by the end of 1945 outlining the possibility of thermonuclear bombs and initiated actual work on them in June 1946. A few years later, it came out that Soviet spies had conveyed detailed blueprints of the American atomic bombs to the Soviets at practically the same time they were being worked out in the United States. This much the Russians admitted after the political changes and the collapse of the Soviet Union in the early 1990s. What may only be suspected, but has not been proved, is that Soviet intelligence had also captured and transmitted information about the thermonuclear preparations as well. But there can be no doubt that the Soviets were already working on their own nuclear weapons, including the hydrogen bomb, at the same time the American scientists were engaged in their fierce debate about the course the United States should be taking.

What was known to the Americans was that very soon after the end of the war, in late summer of 1949, the Soviets exploded their first atomic device. However, this surprised only those who had underestimated the Soviets' preparedness. The extent of Soviet spying was also revealed during the subsequent months and years after the war. The Cold War divided the world for decades; the situation was sometimes made more dangerous by local hot wars of broad significance. Under the circumstances, President Truman had no choice but to instruct the US weapons development community to develop the hydrogen bomb. He declared his decision for the first time on January 31, 1950. The announcement was eventually followed by additional decisions taken in the same direction.

The situation might have resembled the decision about the Manhattan Project, but there were important differences. Among them was the fact that the single and well-defined goal of the Manhattan Project was the creation of the atomic bomb. In 1950, Los Alamos was already engaged in perfecting the atomic bomb and expanding its program considerably. There could not be an all-encompassing program for the hydrogen bomb as there had been for the atomic bomb. In addition, the enthusiasm of many of the potential participants was dampened in part by their initial opposition to the development of the hydrogen bomb and in part by their other obligations. Some of the most important physicists could work for the program for only limited periods. There was no world war going on, and while the threat of the Soviet Union was real, the situation was not the same as it had been between 1942 and 1945.

Perhaps the most conspicuous difference between the two programs was how each had begun: the scientists knew how to build
an atomic bomb at the inception of the Manhattan Project, whereas
nobody knew how to build a hydrogen bomb at the beginning of
1950. This was a bitter realization during the debate over whether
or not to develop the bomb, and it fell on those involved like a cold
shower once the presidential decision to develop it had been made.

The lack of knowledge of how to build the hydrogen bomb
weighed most heavily on Teller, as he had been the most vocal advocate for its development. Now there was the decision to go ahead,
but nobody knew how to proceed. An added difficulty was the lack
of enthusiasm among the Los Alamos leaders toward the new program, because they were already responsible for everything else.
The other programs that stood on solid ground might suffer from a
venture whose outcome was uncertain, to say the least. The difficulties were compounded by the personal animosities that had developed between Teller and many of his peers with whom he was supposed to be toiling jointly. A few very important friends, including
Enrico Fermi and John von Neumann, supported Teller, but their
findings did not support his claims either. There was truth in what
Teller told an interviewer in 1990 about this period: "[I]f I claim
credit for anything, I think I should not claim credit for knowledge,
but for courage. It was not easy to contradict the great majority of
the scientists, who were my only friends in a new country."[2]

Some work on the thermonuclear weapons was going on in Los
Alamos even before President Truman's decision; its proponents,
including Teller, tried to downplay it, while its opponents allegedly
exaggerated it. At this point, Teller might have stopped his lobbying
efforts and focused on the work, but because of the reluctance of
the Los Alamos leadership to divert sufficient support to the project—at least this is how Teller saw it—he continued his lobbying.
Many scientists joined the project, but others elected to stay away

from it. Part of their reluctance stemmed from the worsening internal political atmosphere in the United States. During this period, known as the McCarthy era, Senator Joseph McCarthy indiscriminately cast a shadow of communist tendencies on many, not only among leading intellectuals of the American cultural life but also in such bastions of the American establishment as the state department and eventually the army. The latter would contribute to his downfall, but for a few long years, he created an atmosphere of distrust and unfounded accusations.

The main problem in the apparent lack of perspective on the hydrogen bomb program continued to be that there was no feasible plan for how to build it. There was the idea of the so-called classical Super, according to which an atomic bomb would create the necessary high temperature for the initiation of the fusion reaction in a huge volume of deuterium or mixture of deuterium and tritium; its burning would then propagate throughout the entire volume, producing a tremendous explosion. It did not seem to be workable, though. At this point, had Teller been sure that the hydrogen bomb would be impossible to build, it would have meant that nobody would have it. However, it was tormenting to think that while the Americans might not be able to produce it, the Soviets might be capable of it.

It is to Teller's credit that under such an impossible tension he was still able to think about new solutions. Physicist Luis Alvarez was amazed by Teller's determination to carry on with the work: "Edward Teller told everyone for years that he was going to make a hydrogen bomb. His Super proved unworkable, but since he had committed himself he couldn't walk away. . . . Public commitment is often an essential driving force in invention."[3] The word *frustration* still best characterized the atmosphere of this period. When much work but little real progress was reported at the September 10, 1950, GAC meeting, Oppenheimer expressed "frustrated gratitude" to the participants of the project.[4]

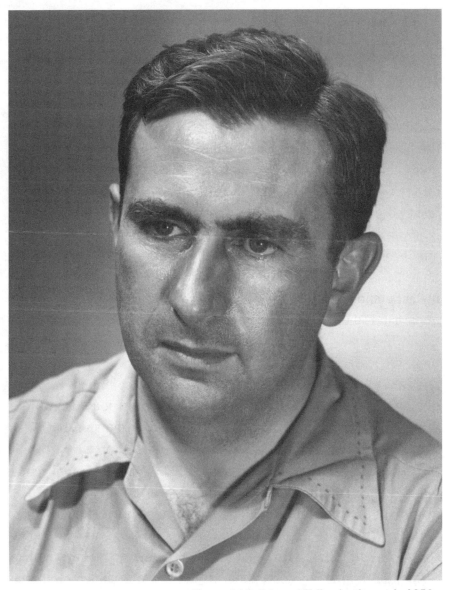

Figure 14.2. Edward Teller in the early 1950s.
Courtesy of Los Alamos National Laboratory.

The classical Super was a relatively simple thing to calculate, because if there was a steady-state burning wave, then it would be essentially stationary in the frame that moves with constant velocity. So the problem was to find a steady-state solution rather than a time-dependent solution. It was still complicated, because there were competing effects, the outcome of which would determine whether or not this idea would work. The higher the temperature, the faster the burning and the fusion of deuterium to deuterium would be. At that time, it was known that the D + D reaction had two channels; one, to make helium-3 plus a neutron, and the other, to make tritium plus a proton. The tritium then also reacts, and its reaction has about a one hundred times higher cross section than the deuterium plus deuterium reaction. Tritium would burn instantly and would add considerably to the energy release. On the other hand, there is also a high-energy neutron in the tritium reaction, which is likely to escape and not contribute to the heating. All these processes required careful calculations.

The scientists knew what they wanted to accomplish; they just did not know how to do it. The only example they could learn from at the time was the thermonuclear reactions in the sun, but these reactions were of limited value to them. Under terrestrial conditions, even higher temperatures are needed than those in the center of the sun, which is estimated to be sixteen million degrees. The process of thermonuclear reaction in the sun is relatively slow, fortunately for us, so that the sun is expected to serve as our energy source for ten billion years.[5] At the much higher temperatures necessary for maintaining thermonuclear reactions under terrestrial conditions, much of the energy is in the form of electromagnetic radiation, which may escape, causing the reaction to die. This was the main problem with the classical Super.

No set of parameters had been proposed that would make the classical Super work, but there was no theorem precluding that it might work. Teller maintained to the end that it might have worked. The tension under which the scientists worked showed its consequences increasingly. Visible fault lines developed between Teller and his friends, such as John Wheeler and von Neumann, in one group, and the leadership of Los Alamos, in the other. Teller continued his lobbying activities, which did not go down well with his opponents. There was little hope of getting unambiguous answers to the questions about the classical Super from theory, hence experiments were planned for the spring of 1951; the series of tests was labeled "Greenhouse." The scientists continued devising these tests to see how to produce thermonuclear reactions, even if they were not necessarily yet suitable for weapon design. The best-known test of the series, called "George," was executed at the Enewetak Atoll in the Pacific Ocean on May 8, 1951. It was a success, showing the presence of thermonuclear reaction for the first time.

By the time the "George" test happened, some important changes had taken place in the work on the program. The concept of the classical Super had become obsolete and had been overtaken by a brand-new idea, if not a new design. As we have seen above, one of the problems with the classical Super was that a lot of energy during the fission explosion would be dissipated by radiation. The higher the temperature, the more probable fusion becomes, but at the same time, more of the energy leaves the system. Taking all these conditions into account, Teller suddenly realized that the particles carry energy in proportion to their number, but radiation carries energy in proportion to its volume. If the number of particles could be increased with a simultaneous decrease of the volume, more energy would stay with the particles. So the question was whether it would be possible to compress the mass of deuterium? Even heavy metals could be compressed in the implosion process, so surely it should be possible to compress liquid deuterium. If the deuterium mass is strongly compressed, thermonuclear reaction would be possible. This is how Teller remembered the events.[6] According to others, however, when the question of compression had come up in discussions, Teller always rejected it.

It is true that the turning point in the program may be ascribed to an idea from the Polish-American mathematician and Los Alamos veteran Stanislaw Ulam. Ulam had been involved in lengthy and painstaking feasibility calculations aimed at deciding whether or not the classical Super might work. He did not worm his way into Teller's heart by repeatedly demonstrating the failure of the concept of the classical Super. When Ulam came to a new idea, it was not Teller to whom he first talked about it but several others who held important positions in the laboratory, though they were much less intimately involved with the Super than Teller. Ulam came up with a practical and feasible solution to the problem. He suggested that instead of the heat produced by the primary fission bomb, they should use the energy in the expanding mass of the bomb's core; in other words, use compression by shock wave. When Ulam finally did tell Teller about his brainchild—how to compress the fusion fuel—Teller's reaction, described here by Ulam, was the following: "At once Edward took up my suggestions, hesitantly at first, but enthusiastically after a few hours. He had seen not only the novel elements, but had found a parallel version, an alternative to what I had said, perhaps more convenient and generalized."[7] Teller told Ulam that he had thought of something that would work even better.

Teller's idea was that they should use the X-rays emitted by the primary fission bomb to compress the deuterium; that is, the compression should happen through radiation. There were at least some elements of the implosion technique of the plutonium-based fission bomb that reappeared in the new approach. One might wonder why nobody had thought of the new solution before. It is not uncommon, though, to see a similar approach being reinvented again and again in different applications as if it had been a completely new idea. As for the question of priority, it makes it even more difficult to assign it when the new approach may have been around "in the air" for some time.

We are not going to analyze the relative merits of Ulam's and Teller's suggestions nor the question of the priority issue. The fact is that Teller found the solution that then proved workable and that moved the hydrogen bomb project from a state of frustration to a complete success. It is also true that Ulam's contribution seems to have played at least the role of a catalyst in Teller's thinking, whether or not Teller was willing to admit this. It is also a fact that Teller often thought most creatively under very strained conditions and came to solutions that were labeled important discoveries.

The documented records provide very little help in getting a more complete picture of this discovery. Teller and Ulam first wrote a sketch of their proposal, and after some further changes, they wrote a joint report, dated March 9, 1951. It was titled "On Heterocatalytic Detonations I: Hydrodynamic Lenses and Radiation Mirrors" and is usually referred to as the "Teller–Ulam design." The document is still classified, and it is doubtful whether its declassification will ever resolve the question of proper assignment of credit between the two scientists. There might have been a chance that some aspect of the controversy would one day be resolved, if each had presented his own contribution in the joint report. However, it is more probable that Ulam's and Teller's contributions are not delineated in their joint report, which was written, in fact, by Teller alone.[8] Teller's idea of authorship and coauthorship was not so much an expression of who originated or worked out the results in an investigation, but he considered it an expression of responsibility.

Toward the end of March 1951, there was a second report; it described a detailed mathematical analysis performed by Frederick

Figure 14.3. (left): Partial view of the "Mike" thermonuclear device. Courtesy of Los Alamos National Laboratory. Figure 14.4. (below): The "Mike" thermonuclear explosion on November 1, 1952. Courtesy of Los Alamos National Laboratory.

de Hoffmann at Teller's suggestion. This analysis placed a second fission component, a subcritical amount of uranium-235, at the core of the thermonuclear material. The analysis fully confirmed what Teller had hoped to achieve. For all practical purposes, it was a joint work between Teller and de Hoffmann, but de Hoffmann put Teller's name on it as the sole author.[9] Teller never forgave himself for not declining this gesture and not insisting on adding de Hoffmann's name to the report.

Curiously, some of the physicists who had been involved with the project of the hydrogen bomb belittled the significance of Teller's discovery, yet gave no explanation why they themselves had not advanced such a solution. It would be incomplete to try to evaluate the value of Teller's discovery without considering at least some of the circumstances that might have influenced the judgment of Teller's peers concerning his contribution to the thermonuclear project. The timing is also important. Teller's pivotal suggestion for building the hydrogen bomb was made in the spring of 1951, and the first successful test based on the new principle, which was designed by the young physicist Richard Garwin, was executed on November 1, 1952. Within a few years' time, two additional events took place that must have colored most everything that people thought and said about Teller.

Teller's dissatisfaction with the leadership of Los Alamos reached such a level during the second half of 1951 that he resigned from the laboratory. For some time, he had started playing with the idea that the United States should have a second weapons laboratory in addition to Los Alamos. His departure intensified his desire to lobby for this idea, and sometime in 1952, the second laboratory, now known as the Lawrence Livermore National Laboratory, opened in Livermore, California. Ernest Lawrence was an enthusiastic supporter of Teller's suggestion, and the second laboratory started its existence as an annex to Lawrence's Berkeley Laboratory, which today is the Lawrence Berkeley National Laboratory. The opening of the second laboratory had an additional divisive effect in the community of physicists; the Los Alamos people feared that Livermore would deplete the ranks of physicists of Los Alamos, though this did not happen. After initial failures, Livermore has become another important venue for the creation of the modern arsenal of weapons for America's defense. After Lawrence's early death, Teller dominated Livermore with his superior talents and authoritarian ways of operation.

The single most damaging event to Teller's relationship with his colleagues was his testimony in the Oppenheimer security hearing in the spring of 1954. Oppenheimer was a national hero in the United States after World War II and a sought-after adviser to the American administrations during the immediate postwar years. His opposition to the development of the hydrogen bomb, however, invited the wrath of the proponents of the thermonuclear program, and he made personal enemies as well, due to his superior intellect and abrasive style. He was not averse to humiliating his adversaries, and he had shown weaknesses of character in his treatment of some of his friends and pupils. In addition, his leftist political past was catching up with him in the atmosphere of McCarthyism in the early 1950s. When his security clearance came into question, the Atomic Energy Commission organized a hearing in 1954 before a Personal Security Board where the usually brilliant Oppenheimer collapsed under some of the accusations against him. He was already like a wounded lion when Teller gave a negative testimony against him—something his peers would never forgive. He was not the only one whose testimony was critical of Oppenheimer, and some argue that what he said was not far from reality, but he was the best known among Oppenheimer's accusers and managed to formulate his words so eloquently that they would be quoted extensively during the subsequent years. Many felt he had betrayed

Oppenheimer, and his standing among the community of physicists never recovered from this self-inflicted damage.

Nonetheless, this negative reaction should not have changed how others judged the importance of Teller's discovery of the solution for thermonuclear explosion. In fact, Hans Bethe, who adamantly and long opposed the development of the hydrogen bomb and agreed to work on it only when it became inevitable—but who was also one of the greatest authorities in the field of thermonuclear physics—evaluated Teller's performance. Rarely has a greater authority in his own field praised more eloquently the discovery of a colleague than Bethe did here with respect to Teller:

> [T]he new concept . . . was entirely unexpected from the previous development. . . . The new concept was to me, who had been rather closely associated with the program, about as surprising as the discovery of fission had been to physicists in 1939 . . . the new concept had created an entirely new technical situation. Such miracles incidentally do happen occasionally in scientific history but it would be folly to count on their occurrence. One of the dangerous consequences of the H-bomb history may well be that government administrators, and perhaps some scientists, too, will imagine that similar miracles should be expected in other developments. . . . Everybody recognizes that Teller more than anyone else contributed ideas at every stage of the H-bomb program.[10]

In a curious twist, anti-Teller sentiments have been manifested in accusations that Teller delayed the development of the American hydrogen bomb by sticking too long to the idea of the classical Super. The basis for such a view might have its roots in the fact that in April 1946 there was a meeting at Los Alamos reviewing the state of affairs of the thermonuclear bomb. The document about the meeting is still classified, but the idea of radiation implosion may (or may not) have been floating around at the time. If it was, it may have originated from Teller or from others. In any case, if the idea had been mentioned, it did not take. But, again, if the idea had come up, it was not only Teller who overlooked it but all the others who were involved with the project. And while nobody seems to be able to answer the question of why it was overlooked, if it was, this question seems to persist.

There are other indications that the idea of radiation implosion might have been around prior to 1951 and that Teller might have heard about it. This would subtract nothing from the importance and uniqueness of his discovery in 1951. In a letter to Maria Goeppert Mayer in mid-winter 1949, Teller writes about his consideration of the hydrogen star. He correlates the possible temperature in its center with its size and finds the temperature estimate realistic "for a star which has stopped growing because radiation pressure counterbalances gravity."[11] This is very revealing and shows that the idea of radiation pressure must have been on Teller's mind at least as early as 1949.

Some of my own encounters with participants in the US nuclear weapons program seemed to indicate that I might have touched a sensitive point by inquiring about the 1946 meeting on the Super. This was surprising to me, since I was not asking about and was not even interested in the technical aspects of the question.[12] Then there are the records of the group interview with Teller at Los Alamos in 1993, during which some penetrating questions were raised about the slowness of the emergence of the radiation implosion concept.[13] These questions were also interesting from the point of view of understanding the nature of scientific discoveries.

Often ideas that seem trivial later appear as shocking novelties when somebody pronounces them for the first time. Also, some ideas that are eventually found to be revolutionary may have been floating around for quite some time, their utility unrecognized by many knowledgeable people. The concept of implosion had already been employed in the plutonium bomb, yet even for Teller, apparently, an additional spark was needed to apply this concept to the hydrogen bomb. This spark was, by all indications, Ulam's original proposal. The question of whether or not radiation implosion was discussed at the 1946 meeting has an additional unsettling aspect. The Soviet spy Klaus Fuchs was present at the 1946 meeting and might have informed the Soviets about radiation implosion. In that case, the solution for the Soviet hydrogen bomb might not have been their genuine invention, as the Russians have claimed with great emphasis.

Another idea whose inception would be interesting to scrutinize further is the application of lithium(6) deuteride as fusion fuel. The idea of using lithium as fusion fuel came about for the first time in a letter from Teller to Oppenheimer in 1942. Thus it had been around in plenty of time for Fuchs to pick it up, so he might have conveyed this idea to the Soviets as well. In the Soviet program, it had been recorded that the later Nobel laureate Vitaly L. Ginzburg had suggested using lithium(6) deuteride. There is no doubt as to the originality of Ginzburg's suggestion because—lacking security clearance—he could not have had access to products of espionage, whereas the other Soviet scientists in the program might have had such access. Indeed, Ginzburg did not even learn that lithium(6) deuteride was used in the Soviet program until the collapse of the Soviet Union. It is not known, however, whether the use of lithium(6) deuteride by the Soviets was based on Ginzburg's suggestion or on prior intelligence. The Americans did not use lithium in their thermonuclear devices until after the "Mike" test; that is, a little more than ten years after Teller's letter to Oppenheimer in 1942.

The hydrogen bomb was (and continues to be) a most horrible weapon, a shameful development in the history of humankind. Some say that its development has proved unnecessary because, in spite of its tremendous expense, it was never used. This, however, must be countered by the fact that because both superpowers possessed it, it was never used. Had Stalin (and his successors) unilaterally become the sole possessor of the hydrogen bomb, what guarantees were there that he would not have blackmailed the Free World by using it as a threat?

Teller has been accused of having forced his agenda onto the United States. However, it would almost be ascribing him supernatural power to suppose that he could have singlehandedly imposed his will upon the Truman administration and personally upon President Truman. In an ideal world with a proper system of interactions between government and science, Teller's lobbying might have been rendered superfluous. However, he remembered a situation merely one decade before, when a small group of immigrants took it upon themselves to warn the president of the United States of the possible danger of an atomic bomb of heretofore unimaginable strength that might be coming into the hands of Nazi Germany. Nobody accused Leo Szilard of exaggerating the dangers, though he was accused of inefficiency in initiating the development of the atomic bomb.[14] Teller felt the torment of facing a yet stronger adversary of the United States that once again might have the opportunity to acquire a thousand-times-more-powerful weapon than the atomic bomb while the democracies stood idly by.

CHAPTER 15

JOY OF UNDERSTANDING
Big Bang

[Gamow] went wherever his curiosity led.
Vera C. Rubin[1]

Gamow was fantastic in his ideas. He was right, he was wrong. More often wrong than right. Always interesting.
Edward Teller[2]

George Gamow (1904–1968) found sheer joy in understanding our world, and this drove him to thoughts that others found unorthodox. He worked out a theory of the origin of the universe, which his critics ridiculed by calling it the Big Bang. However, when independent discoveries supported the model, Big Bang *became a respected term. Some considered the understanding of the origin of the universe to be one of the most important discoveries of all time. This discovery was preceded by Gamow's outstanding contributions to nuclear physics and was followed by another original idea in the fledgling field of molecular biology. In addition to his seminal research, Gamow wrote dozens of well-received popular-science books that are still in print and help a broad readership understand modern physics. Gamow scarcely received any recognition in his life, yet he was graciously content with the ideas he advanced, the knowledge he gathered, and the possibility of sharing it all with others.*

"The problems of cosmogony—that is, the theory of the origin of the world—have perplexed the human mind ever since the dawn of history."[3] With these words, George Gamow introduced his book *The Creation of the Universe* in 1952. The word "creation" in the title caused some consternation among the reviewers, so the author clarified it in the second printing of the book. He did not mean "making something out of nothing";[4] rather, he meant it in the sense of, for example, "the latest creation of Parisian fashion."[5] Gamow's name has been associated with the idea called the Big Bang, which is today the most accepted theory about how the universe began.

George Gamow was born in Odessa (then Russia, now the Ukraine) into a family of teachers. His birth name was Georgii Antonovich Gamow, and he usually signed his name as "Geo" pronounced as "Joe." He started higher education in his home town and then moved to Petrograd State University, which was soon renamed Leningrad State University, and which is today known by its yet older name of Saint Petersburg State University. Parallel to his studies, he held several jobs, among them an instructor's position at an artillery school of the Red Army, where he carried the remuneration and rank of a colonel.

While he was a student, Gamow was introduced to the theory of relativity by the Russian mathematician Alexander A.

Figure 15.1. George Gamow (top row, first from the left), Dmitri Ivanenko (top row, second from the right), and Lev Landau (front row, second from the right) with fellow students in Leningrad, 1926. Photo courtesy of G. A. Sardanashvily, Moscow.

Friedmann, who became his favorite teacher. Friedmann had created a theory of the expanding universe, which contradicted Einstein's views of cosmology, and which Einstein first refused but later accepted. Friedmann had kindled Gamow's interest in theoretical physics, especially in nuclear physics, before he died prematurely in 1925. Friedmann died without realizing that his two brief papers would have considerable impact on the development of modern cosmology.[6]

In Leningrad, Gamow's friends were his fellow students: Dmitri Ivanenko, Lev Landau, and Matvei Bronshtein. Years later, under Stalin's repressive regime, Ivanenko and Landau spent about one year each incarcerated, and Bronshtein was shot after having been arrested on fabricated charges. Gamow was the only one among them who would not have to endure such an experience; probably only because by then he was no longer in the Soviet Union. Eventually, both Landau and Ivanenko became superstars in Soviet physics. According to Ivanenko, Gamow was tall, strong-built, with regular facial features. He had a high-pitched voice and was myopic, but he started wearing glasses only during his university years. He drew well, was active in conversation, and loved practical jokes. He showed hardly any interest in politics and did not read newspapers.[7]

Gamow got off to a fast start in science and was only twenty years old when, in 1924, he gave a talk at a congress of Russian physicists in Leningrad. His first paper was published at the age of twenty-two; it was a joint article with Ivanenko about the wave mechanics of matter. Another paper by Gamow, Ivanenko, and Landau, in 1928, was of a very general character about universal constants—Planck's constant, h, the speed of light in vacuum, c, and the constant of universal gravity, G—and limiting transitions.[8] When in a little more than a dozen years Gamow would start writing his popular-science books, he chose these universal constants for the initials of his imaginary hero, c. G. h. Tompkins.

After receiving his master's degree, Gamow started his doctoral studies, but he interrupted them in the summer of 1928 when he was sent to Göttingen in Germany, one of the world centers of physics at that time. There, he joined one of the most advanced groups of theoretical physicists in the world. At its center was Max Born, who regularly held his prestigious seminars in the Institute of Theoretical Physics at the University of Göttingen. Gamow observed the spectacular development of quantum mechanics, but the new field attracted so many researchers that he did not want to become just one of the many involved. He preferred to stick to nuclear physics, a much less crowded field at that time.

Gamow keenly followed the developments of this field wherever they took place. He was very impressed by Ernest Rutherford's discoveries in Cambridge, United Kingdom, of the structure of radioactive atoms and the origin of alpha particles. An alpha particle is a helium nucleus (atomic number 2)—when the helium atom is stripped of its two electrons. Rutherford studied the nuclear reaction when uranium (atomic number 92) emits an alpha particle and transforms into thorium (atomic number 90), the element whose atomic number is two units smaller than uranium's. The atomic number corresponds to the number of protons in the nucleus. The following equation indicates the changes in the atomic masses, which correspond to the sum of neutrons and protons in the nuclei:

$$^{238}U \rightarrow {}^{234}Th + {}^{4}He^{2+}$$

where $^{4}He^{2+}$ is equivalent to an alpha particle.

It was puzzling that the alpha particles were not supposed to have sufficiently high energy to leave the nucleus, but they did. Rutherford described the phenomenon but did not provide an interpretation for it. Gamow found an explanation for the phenomenon, subsequently called a tunneling effect, by using the new theory of quantum mechanics and showing that alpha decay was possible at lower energies than the presumed energy barrier. Gamow did not stop at providing a theoretical explanation but suggested an experiment in which accelerated protons—possessing less energy than they might be thought to need to penetrate the coulombic (that is, purely electrostatic) barrier—would be able to split light atoms like lithium.

For Gamow, the real significance of his involvement in this question was that he made an impact on his European colleagues, which paved his way to Rutherford's Cavendish Laboratory. His ideas prompted Rutherford to instruct his associate John Cockcroft and his student Ernest Walton to carry out the experiment in 1932 for which they would receive the Nobel Prize in 1951.

Within a short time, Gamow's scientific acumen was recognized by such greats as Born, Niels Bohr, and Rutherford. As a result, he extended his visit in western Europe and stayed for some time in Copenhagen and Cambridge. Upon Gamow's return to the Soviet Union in the spring of 1929, he and his results were praised in the Soviet media as the achievements of the new order. The famous Soviet poet Demyan Bedny published a poem in the party newspaper *Pravda* honoring the young physicist. Gamow did not stay home for long, and in the fall of 1929, he was already back at Cambridge on a Rockefeller fellowship. He continued his studies in nuclear physics, wrote reviews for a Soviet journal, and produced his first book when he was twenty-six about the atomic nucleus and radioactivity.

Gamow traveled a lot, made friends, and became well-known and popular among physicists, especially theoreticians. He returned to the Soviet Union in the spring of 1931. This time, Gamow found a hardened political atmosphere, which was very different from that of previous years. In the 1920s, the Soviet government encouraged the leading researchers as well as gifted budding scientists to strengthen interactions with the West and helped them to visit lab-

oratories there. This changed almost abruptly with the consolida-
tion of Stalin's power over the country. The Soviet leadership now
looked at science as a weapon in the fight against capitalism, while
it was becoming pathologically apprehensive of Western ideas penetrating into the Soviet Union.

Gamow's next two years in the Soviet Union were full of frustration. He could hardly do any science, in great contrast with his
fruitful activities in western Europe during the preceding years. He
continued receiving invitations to attend meetings in the West, but
he was not permitted to travel. His troubles, though, brought him a
side benefit. On one of his visits to the passport office, he met a
physics student by the name of Lyubov' N. Vokhmintseva, or Rho,
as Gamow eventually called her. They fell in love and married soon
after meeting. From this point on, Gamow was not alone in his
scheming to leave the Soviet Union. The couple made adventurous
attempts to escape, but without success. Gamow and Rho were
married for twenty-five years before divorcing in 1956. They had
one son, Rustem Igor, born in 1935, in the United States.

Apart from his frustrations, Gamow lived under better conditions than most in the Soviet Union. He had several jobs and was
recognized as one of the leading figures of Soviet theoretical
physics, in spite of his young age. He was the first among his peers
to be elected to the Academy of Sciences, as corresponding
member, in the spring of 1932. There were also failures, to be sure:
Gamow, Landau, and a few others made an unsuccessful attempt to
organize a separate institute of theoretical physics in Leningrad.

Finally, in 1933, Gamow and his wife succeeded in getting permission to leave the Soviet Union, thanks to a prestigious invitation
to Gamow to attend the next Solvay meeting of physicists in Brussels. The meeting brought together an exceptional collection of
leading international physicists, but for Gamow, the gathering was
more than a chance to be among these luminaries; it was his

gateway to freedom. Soon gossip started circulating in Moscow that he and his wife would not be returning to the Soviet Union. Leading figures in the international community had aided Gamow in getting out of the Soviet Union, but they did not think he was leaving for good—especially not illegally. Bohr was involved, and so were Marie Curie and even Paul Langevin, who was not only politically sympathetic toward the Soviet Union but was in charge of the French-Soviet friendship organization.

When after the Brussels meeting Gamow declared that he had no intention of returning to the Soviet Union, his sponsors arranged for the Gamows to stay for as long as possible legally; that is, they had their permission extended before formally declaring defection. After his defection, Gamow became a nonperson in the Soviet Union, and for decades his name could not be pronounced at official meetings, even though the leading physicists continued to follow his scientific activities. In the 1980s, many years after he had passed away, Gamow's works began getting due recognition, and in 1990, his membership in the Soviet Academy of Sciences was reinstated.[9]

As for Gamow's experiences in the West, he worked some temporary jobs, then got the opportunity for long-lasting employment in the form of an attractive offer from George Washington University (GWU). The ambitious president of GWU, Cloyd H. Marvin, wanted to expand the university and elevate its academic level. GWU had an outstanding law school and a similarly prestigious medical school but did not excel in science. Marvin established a fund for creating a strong Physics Department, but even the then-large sum of $100,000 he amassed would not suffice to fund a department strong in experimental physics. Rather, he had been advised to build up a powerful group in theoretical physics for which only scientists, travel money, paper, and pencil would be needed.

Thus Marvin invited Gamow to GWU in the fall of 1934 and complied with Gamow's two stipulations. One was that he needed a partner, and so Edward Teller was invited in this capacity. The two scientists augmented each other both by temperament and by their areas of expertise, and their interactions resulted in excellent joint research. Gamow's other stipulation was that he wanted an annual meeting of theoretical physics to help build up the field in the United States—a field that had not flourished heretofore in this country. Gamow and Teller organized these meetings with great success.

Gamow's fame in the early 1930s came from his discoveries in nuclear physics, but he never abandoned his interest in astrophysics and in the origin of the universe—an interest that was inculcated in him by his former teacher Alexander Friedmann. Gamow realized that these two domains of science coalesced and that his experience with nuclear physics would help him make strides in cosmology. We need to mention yet another pioneer, Arthur Eddington, who in 1926 advanced his suggestion about the nuclear nature of the energy production in the stars. It was the reaction of hydrogen turning into helium,

$$4H \rightarrow He + \Delta E,$$

where the amount of energy (ΔE) corresponds to the mass difference, also called mass defect (Δm), between one helium atom and the four hydrogen atoms. This is according to Einstein's famous equation $\Delta E = \Delta mc^2$. This nuclear reaction can take place only at very high temperatures that enable the hydrogen atoms to overcome the electrostatic repulsions between them, hence the name, thermonuclear reaction. It is also called *fusion* because it involves the joining of atoms (in contrast with *fission*, in which a heavy nucleus splits into two nuclei). Eddington's theory suffered from the fact that the estimated temperature in the interior of the stars, such as the sun, although very, very high, was still estimated not to be high enough for the reaction.[10]

Gamow's interest in thermonuclear reactions dated back to his life in the Soviet Union. He used to give lectures to broad audiences on thermonuclear reactions and earned some fame as a consequence. One time, Nikolai Bukharin, a high political operative, attended Gamow's lecture, after which the politician offered Gamow the opportunity to branch out in the direction of experiments and to set up a thermonuclear program. He offered Gamow the use for one night a week of all the electrical power of the Moscow industrial district for his experiments. Gamow declined the proposal.

In 1935, the Ohio chapter of the Scientific Research Society Sigma Xi organized a symposium "The Nucleus of the Atom and Its Structure" with Gamow's participation. There were five lectures—one of the others was by Ernest O. Lawrence—and they were published in the rather obscure *Ohio Journal of Science*. In his lecture, Gamow pointed out the importance of neutron capture in forming new elements and the role of neutrons in the energy production in the stars.[11] The neutron had only recently (in 1932) been discovered.

Both Gamow and Teller were interested in the nuclear processes in the stars when they interacted at GWU during the second half of the 1930s. This shared interest gave them the idea of organizing their conference in 1938 about the energy production in the stars. They urged their friend Hans Bethe to attend the 1938 meeting. Bethe was an expert in nuclear physics, but his interest in cosmology was kindled during the Washington conference. In 1967, Bethe was awarded the Nobel Prize in Physics primarily for his discoveries concerning the energy production in the stars.

Bethe developed his theories of how nuclei were synthesized, but he also realized that the conditions in the sun were colder than what was needed for the reactions he envisioned. Bethe, together with Gamow and Charles Critchfield, realized the basic importance of the proton–proton reaction in the sun, and this reaction corresponded to a previous piece of research by Gamow and Teller, which established the temperature-dependence of thermonuclear reactions. This was Teller's first involvement with the topic that would eventually play a central role in his projects for developing the most powerful weaponry (chapter 14).

The 1938 Washington conference might have caused Gamow to return to basic problems related to his youthful interest, but World War II diverted his attention, and he became busy with war-related research for the navy. It might have been more natural for him to join the Manhattan Project, but he was not cleared for participation because of his ostensibly high rank in the Red Army; in reality, he had merely held an instructor's job. Eventually, he became a participant in the efforts to develop the hydrogen bomb after the war, but he did not play any conspicuous role.

When Gamow resumed his work at GWU in 1946, he finally combined his research in cosmology and nuclear physics. He already had a graduate student, Ralph Alpher, working on the origin of the universe. Gamow's concern for the understanding of the formation and distribution of the elements prompted Gamow and Alpher to direct Alpher's dissertation work toward helping scientists understand the synthesis of the chemical elements in the early phase of the universe. This early universe was supposed to be hot, dense, and expanding. Gamow and Alpher used a mathematical model based on the teachings of Friedmann and another pioneering scientist, Georges Lamaître. They used all available information about nuclear reactions and introduced many reasonable assumptions to simplify their calculations. They still had to carry out an enormous amount of calculations, and at that time, there were no fast computers to aid their work. In the end, the team produced estimates of the relative abundance of the chemical elements in the universe, which pointed to a high relative abundance of helium, which added to the credibility of their results. It was a big success for the

Big Bang theory, because the distribution of the light elements was known from entirely independent sources.[12]

The paper in which they reported their findings has an interesting history. Gamow and Alpher wrote it on the basis of Alpher's dissertation, to which Gamow—ever the great joker—added Bethe's name for euphony. He claimed that it was "unfair to the Greek alphabet to have the article signed by Alpher and Gamow only."[13] Thus, the paper was published by Alpher, Bethe, and Gamow.[14] That the paper was published on April 1 did not help it to be taken seriously, but that was not done by design. The paper was a step toward a description of the universe, which is often referred to as the alpha–beta–gamma model. Curiously, Bethe gave his consent to allowing his name to appear on the paper without having contributed to it.

The resulting report was the first serious pronouncement of the Big Bang model, though it had not yet been given this name. The name *Big Bang* was used later in criticism of the model; thus it was not originally used in a complimentary way. But the name stuck and became respectable with the later acceptance of the model. Initially, however, the Big Bang model had a difficult time getting recognized. The prevailing view supported the so-called steady-state model, whose main proponent was the noted British astronomer Frederick Hoyle (who, incidentally, coined the name *Big Bang*). Nonetheless, Alpher had no difficulty in getting his PhD degree on the basis of his work. He defended his doctoral dissertation in front of an unusually large audience of three hundred spectators.

When Gamow first suggested adding Bethe's name to their paper, Alpher was rather apprehensive about it, in case his name would not be recognized in the shadow of the two well-known names. His apprehension was not without foundation, and subsequent events showed that he was right to be concerned. For quite a while, Alpher's contribution, along with his colleague Robert

Herman's contribution—with whom he had continued the work—almost disappeared in oblivion. When the first reports of the observational evidence for the Big Bang theory emerged, nobody seemed to remember Alpher and Herman, including those who produced the new and revolutionary information.

The early versions of the Big Bang model could not provide satisfactory answers to the questions about element formation beyond a rather low atomic number. Alpher and Herman thus kept refining the model by eliminating some of the assumptions, which meant that they had to invest a lot of additional work into their model. This was very different from Gamow's style of work. Vera Rubin, whose PhD work Gamow directed, found him "very pleasant, very amusing, [he] liked scientific games and jokes, but was not really interested in the scientific details of an analysis."[15] Eventually another scientist, John Follin, joined Alpher and Herman, but their work progressed very slowly. They did not even work in the same place and often could only work on the project in their spare time. But, from time to time, great scientists such as Enrico Fermi and Eugene Wigner made some forays into this work, which helped keep it alive.

Much has been described in great detail about the development of the Big Bang model, but its essence could be formulated in a simplified way. According to Gamow, the universe started with an explosion, and the matter that came out of it was extremely hot. For the acceptance of the model, all kinds of problems persisted, including the uncertainty about the age of the universe, which came out to be unrealistic from the early estimates, and the continual problem of the formation of the heavier elements.

A crucial consideration was a consequence of the model that there were remnants of the cosmic microwave radiation—also called remnant heat—which should have stayed around even billions of years after the moment of the Big Bang. Of course, the amount of the cosmic microwave radiation should have diminished from the extremely high temperatures to a few kelvins on the absolute temperature scale. In a tour de force study—not completely understood even by experts—Gamow predicted the temperature of the universe to be seven kelvins. Here is an overly simplified description of his approach: he took two quantities, the age of the universe and the average density of matter in the universe, and from these he estimated the third quantity, the temperature, corresponding to the cosmic microwave radiation. His exceptional insight into the physical problems enabled him to reduce the most complicated problems to a bare minimum. Gamow first communicated his result in another obscure though respectable periodical of the Royal Danish Academy of Sciences.[16] It has to be added that Gamow's elegant estimate followed Alpher and Herman's prediction of five kelvins in 1948, which was the result of an enormous number of painstaking calculations.

When in 1964 Arno Penzias and Robert Wilson of Bell Labs communicated their serendipitous observation of the cosmic microwave radiation amounting to three kelvins, it was a stunning confirmation of the Big Bang model. Penzias and Wilson did not set out to provide proof for the Big Bang; they were not concerned with models about the origin of the universe. They wanted to understand the mechanism by which the galaxy radiates and to illuminate the electrodynamics of the Milky Way. That was their plan, and they started by calibrating their equipment and pointing it at the sky, where they expected nothing at all—or, if there should be some noise, it would be at a very low level. What they observed was about four kelvins, which is not much heat according to our everyday experience, but for them the sky appeared much "hotter" than it should have been. They spent a whole year trying to make sure that what they were observing was not something of an artifact.[17] They checked the possibility that the high level of "noise" might originate from trivial natural sources. For example, they thoroughly scrubbed their antenna to rid it of pigeon droppings that lowered the temperature by half a kelvin, which was far from the total effect.

When Penzias and Wilson finally convinced themselves of the reality of their measurement, they started looking for an explanation. If they were aware of the Big Bang model, they did not make a connection with it. In yet another remarkable coincidence in timing, other scientists at Princeton and elsewhere turned one more time to the possible models of the origin of the universe. Penzias and Wilson learned about the cosmological interpretation of their observation work by the Princeton group, and the two teams published their papers back-to-back in *Physical Review* in 1965. In 1978, Penzias and Wilson shared half of the physics Nobel Prize for their discovery of cosmic microwave background radiation. The other half of the prize went to Peter Kapitza for unrelated achievements.

**Figure 15.2. Robert Wilson and Arno Penzias
in Stockholm, 2001. Photo by the author.**

In his Nobel lecture, Penzias discussed the origin of elements, reviewing all the history needed to understand his and Wilson's discovery. He paid proper tribute to Gamow's and then to Alpher and Herman's contributions.[18] The temperature Penzias and Wilson reported was given as 3.1±1K, meaning that it could be anywhere between two and four kelvins. With such a relatively large uncertainty, Alpher and Herman's five kelvins—and even Gamow's seven kelvins—looked not just respectable but outright excellent, if considering the inherent uncertainties of the circumstances of their estimations. Of course, the main point was that the number was unambiguously different from zero. In other words, it was definitely established that there is a remnant heat in the universe, giving final and absolutely convincing evidence for the Big Bang model. Hoyle was among those who not only welcomed the model that had been his rival but contributed evidence to its recognition.

Penzias and Wilson's observation, and the subsequent triumph of the Big Bang model, made Gamow happy. He was gracious in recognizing the achievements of others, as evidenced in his analogy "I lose a nickel and you find a nickel. I can't prove that it's my nickel except that I lost a nickel just there."[19] This magnanimity, however, did not prevent Gamow from pointing out to Penzias the deficiency of the early history as Penzias had initially presented it. In his letter of September 29, 1963, to Penzias (reproduced here in facsimile), Gamow referred to some of his early theoretical papers in which he mentioned what he called the "primeval fireball" and— significantly—to Alpher and Herman's 1949 paper in which they estimated the remnant temperature as five kelvins. Obviously, Penzias took notice and referred to these contributions in his Nobel lecture (see above).

It is not Gamow's fault that the significance of Alpher and Herman's works have paled in the world literature of science behind his shadow. While Alpher and Herman eloquently expressed

The Sept 29 ᵗʰ 1863
Gamow Dacha
765 - 6th Street
Boulder, Colorado

Dear Dr. Penzias,

Send Thank you for sending me your paper on 3 °K radiation. It is very nicely written except that "early history" is not "quite complete". The theory of, what is now knows as "primeval fireball" was first developed by me in 1946 (Phys Rev. 70, 572, 1946 ; 74, 505, 1948 ; Nature 162, 680, 1948). The prediction of the numerical value of the present (residual)temperature could can be found in Alpher & Hermann's paper (Phys. Rev. 75, 1093, 1949) who estimate it as 5 °K, and no in my paper (Kong.Dansk.Ved.Sels. 27 no10, 1953) with the estimate of 7°K. Even in my popular book "Creation of Universe" (Viking 1952) you can find (on p. 42) the formula T=1.5·10⁻¹⁰/t½ °K, and the upper limit of 50 °K. Thus, you see the world did not start with almighty Dicke. Sincerly G.Gamow.

Figure 15.3. Facsimile of George Gamow's letter of September 29, 1963, to Arno Penzias. Courtesy of Arno Penzias, Menlo Park, California, and by permission from R. Igor Gamow, Boulder, Colorado.

their devotion toward Gamow, they unabashedly described the problems with proper attribution to their results, which was not unique in this case. As they stated, there is "the tendency of scientists to attribute a piece of work with which several names are associated to the author whose name is a household word in the world of science."[20]

Gamow became something of a hero to Penzias, who considered him a greater scientist than Galileo, at least as far as their contributions to cosmology were concerned. Penzias and Wilson were somewhat overwhelmed by the importance of their discovery when it dawned on them. It was significant that the *New York Times* put it on its front page, even before their research paper was published in *Physical Review*.

Gamow's original Big Bang model could be labeled as a premature discovery if we define as premature those discoveries that the rest of the state of science is not yet ripe enough to absorb.

Gamow, obviously, did not see it this way; his genius manifested itself in the fact that he was able to bring together seemingly disparate facts, observations, and theories, and make a far-reaching conclusion from them. His primary synthesis was built equally on his experience in nuclear physics and in cosmology. Where he sensed gaps, he augmented the missing links with intuition and imagination. He was always alert for novelties, whether they were within the realm of his expertise or far from it. He had an insatiable thirst for knowledge in any field of human inquiry, regardless of whether or not he was qualified for the particular field in which he found an intriguing puzzle.

Nothing showed Gamow's interest in a domain of science far from his expertise more conclusively than his reaction to the discovery of the double helix. When James Watson and Francis Crick announced their model of the structure of the molecule of life, DNA, Gamow sensed at once that there must be a process of information transfer from the nucleic acids to proteins. Today, this information transfer is described in the genetic code. A few weeks after

Figure 15.4. George Gamow and a model of the double helix of DNA. Courtesy of R. Igor Gamow, Boulder, Colorado.

Watson and Crick published their second paper on the significance of the double-helix model of the DNA structure, Gamow wrote them a letter. He raised the question of how proteins may come into existence on the basis of the genetic information stored in DNA. He tried unsuccessfully to solve this puzzle, but it was more significant that he focused attention on this crucial question and pushed the researchers in the direction of looking for answers.

The originality of Gamow's thinking did not stop with his first question. He initiated a playful organization called the RNA Tie Club, with *RNA* standing for ribonucleic acid. There were to be twenty regular members—each assigned to one of the twenty naturally occurring amino acids—and four honorary members, each assigned to one of the four bases of DNA. Each member of the club was to wear a customized tie and a tie pin.

The club and all its paraphernalia was an amusing exercise for Gamow, but only a superficial observer would consider the RNA Tie Club to be merely a childish diversion. In reality, it was a modern way of exchanging ideas quickly and without the formalities of official publications among the top researchers in a fast-developing field. Its members found it very useful to be informed about new ideas and experiments even before they reached publishable stages. The way they communicated with each other could only be compared with today's network of scientists working in related areas and using the Internet for fast exchanges without peer review.

The activities of the club peaked in the second half of 1954. Then it fizzled out. Great names such as Melvin Calvin, Erwin Chargaff, Nicholas Metropolis, Max Delbrück, Sydney Brenner, Fritz Lipmann, and Albert Szent-Györgyi were among the members, in addition to Gamow, Crick, and Watson. The world of DNA-RNA scientists was small at the time. Much of its community could be squeezed into this small circle, and there was even space for some physicist friends of Gamow, like Richard Feynman and Edward Teller.

There is yet another area of Gamow's activities that is an essential part of his legacy: his books. His engaging autobiography, *My World Line*, carries the reader through his adventurous life up to about the beginning of his "American" period.[21] Sadly, the book has been out of print for quite some time. However, many of his other books are still available, including his stories popularizing physics of whom the central hero is the Mr. Tompkins mentioned previously. The only award Gamow received during his entire life was for these books, when, in 1956, UNESCO awarded him the Kalinga Prize. As the prestige of a prize greatly depends on its recipients, it is worth mentioning that Julian Huxley (1953), Bertrand Russell (1957), and Fred Hoyle (1967) have been among this stellar group.

George Gamow was a unique scientist and a unique human being. The way he did science was unlike anybody else's approach, and if it was unorthodox in his time, it would be even more so today in the era of big budgets, big equipment, and hundreds of authors' names on one paper. But we can be sure that today's science would also benefit from having another Gamow around.

EPILOGUE
Lessons, But No Recipes

There are no recipes for making scientific discoveries. On the contrary, aiming to make such a recipe is usually a recipe for frustration. In the preceding chapters, we enumerated some general and specific characteristics of a set of successful researchers. Would emulating them lead to success? There is no guarantee of that. Would imitation be useful? It might be, depending on whether or not the object is worthy of imitation.

We have seen fifteen cases of success in science in which the scientists displayed discernible individual traits in addition to drive and curiosity. It was a key to success that these scientists tailored their activities to best utilize their unique talents.

James D. Watson showed that even if one is not the greatest scientist, he or she can still make seminal discoveries. He was brave to set himself a goal others thought premature at best. He found himself a similarly intelligent partner, absorbed all available information by reading, talking, observing, or snatching; he integrated this information and arrived at a discovery that shook science for the next fifty years and whose consequences will affect the world for a long time to come.

Gertrude B. Elion embarked on her research career without a PhD, but she found a gifted boss who was generous with his knowledge. He kept moving up on the corporate ladder and cleared the

297

way for her to move up in his footsteps. She did not mind remaining number two during most of her career. She was dedicated to finding cures.

Both MRI discoverers **Paul Lauterbur** and **Peter Mansfield** had traveled in the slow lane in their respective careers. Both showed stamina and learned all they needed to know from basic science. The gist of their success was that they utilized what they learned in solving something entirely new. They could transform their knowledge (possessed by many) to something they pioneered. This ability to transform knowledge to a different set of conditions is an immensely useful trait both in science and other fields.

Linus Pauling engaged in the transformation of knowledge when he applied the accomplishments of modern physics to chemistry. His achievements made it possible for others to make discoveries that would shake the world of science and medicine.

Frederick Sanger's success paved the way toward the revolutions of medicine in the decades to come. His knowledge was both specific and vast. He was not given to delegate even mundane tasks, and it was easy to underestimate him. But he knew what he knew and made the most of it.

Árpád Furka is another example of stamina combined with talent. He became the recognized pioneer of a new domain in chemistry in spite of a multitude of handicaps throughout his career. He was brave enough to pose fundamental questions and did not think that doing so was beyond his call.

For **Rosalyn Yalow,** recognition meant little during much of her career. This changed when she was left alone and felt the weight of continuing the quest of her partner, Sol Berson. She was tough and aggressive, and on top of her science, she took care of her children and family—something most tough and aggressive men do not have to do—and she prevailed.

Dan Shechtman made a discovery that others had proved

many times over as impossible. He braved the wrath of his peers and was irreverent enough to weather the disapproval of the authorities of his field. He persevered and succeeded.

The discovery of conducting polymers was above all the story of risking a scientific reputation. It was due to **Alan MacDiarmid**'s initiatives that the serendipitous finding of **Hideki Shirakawa** and his Korean associate Hyung Chick Pyun did not disappear in oblivion, and that **Alan Heeger** had an opportunity to apply his knowledge of condensed-phase physics and his entrepreneurial abilities to marketable devices.

F. Sherwood Rowland investigating what happened to the ozone layer under the impact of CFCs is another case of risk taking. It helped that he was a meticulous scientist, as is exemplified by his careful studies of the mercury pollution of fish, which led to exonerating industrial polluters. The discovery made by Rowland and his postdoctoral associate **Mario Molina** was not in fundamental science, but it utilized fundamental science in a discovery for the protection of the environment.

Kary Mullis came to his multibillion-dollar-worth discovery by letting his imagination go. There were more knowledgeable scientists who might have made the same discovery, but they did not; Mullis did. He did not feel confined in his thinking, and his mind was well prepared for his jumping ideas.

Neil Bartlett was on the lookout for new things to make. He relied on the experience of others and was just barely ahead of his more powerful competitors. He charged ahead and produced something nobody had succeeded in doing before.

Leo Szilard has been accused of not carrying through his ideas to mature discoveries. But this was his forte, too, because he could come up with numerous ideas without getting bogged down by the details of a select few. He did not want to do what others could also do. He knew himself sufficiently and what he wanted to achieve.

He pursued his goals despite what others might have expected of him.

Edward Teller could perform even under extreme tension—or perhaps because of it. He often did research at the interphase of different domains of science and moved on after he had solved a problem, choosing not to waste his talent on repeating or working out details. Whereas he alienated his peers with his politics, they recognized his significant discoveries.

George Gamow always did what entertained him. His work was never routine and never what others could have done. He also excelled in producing popular-science books. Both his Big Bang model and his books will be his legacy.

*   *   *

In modern science, the interdependence of various fields has increased, yet crossing disciplinary boundaries has remained difficult. In our education, we are instructed in the framework of disciplines, and such frameworks direct us in obtaining our degrees and in our work in academia. Successful scientists are often the ones who cross these boundaries, and many of the important discoveries in modern science have been made in border areas of the disciplines. The scientists who figured in the preceding chapters often crossed disciplinary borders during their careers.

I have long thought about whether I could single out a general conclusion from the stories in this book that would be sufficiently broad to be applicable not only to scientific research but also to other fields of human endeavor. Of course, each of us may conclude something different from the fifteen cases as determined by our own background and interests. The conclusion I offer is this: In order to accomplish the most, whether in science or elsewhere, *the best one can do is doing what one is best at doing*.

NOTES

PREFACE

1. The book—Péter Teknős, *Az ezerszínű kincs* (The Treasure of a Thousand Colors) (Budapest: Ifjusági Könyvkiadó, 1951)—was the prize for winning a local competition in mathematics when I was in the fifth grade. It was highly politicized in the spirit of the darkest Stalinist period, but this is not what I remembered about it. What caught my imagination was the plethora of useful things it was possible to produce from the most ordinary materials, like coal.

2. Albert Szent-Györgyi, *Bioenergetics* (New York: Academic Press, 1957).

3. István Hargittai with Magdolna Hargittai and Balazs Hargittai, *Candid Science: Conversations with Famous Scientists I–VI* (London: Imperial College Press, 2000–2006).

INTRODUCTION

1. István Hargittai and Magdolna Hargittai, *Candid Science VI: More Conversations with Famous Scientists* (London: Imperial College Press, 2006), "Peter Mansfield," pp. 216–37; actual quote, p. 233.

2. James D. Watson, quoted in M. Cook, *Faces of Science* (New York: W. W. Norton, 2005), p. 160.

CHAPTER 1

1. James D. Watson, "The Human Genome Project: Past, Present, and Future," *Science* 248 (1990): 44–48.

2. James D. Watson and F. H. C. Crick, "Molecular Structure of Nucleic Acids: A Structure for Deoxyribose Nucleic Acid," *Nature* 25 (April 1953): 737–38.

3. Oswald T. Avery, Colin M. MacLeod, and Maclyn McCarty, "Studies on the Chemical Nature of the Substance Inducing Transformation of Pneumococcal Types: Induction of Transformation by a Desoxyribonucleic Acid Fraction Isolated from Pneumococcus Type III," *Journal of Experimental Medicine* 79 (1944): 137–58.

4. Alfred D. Hershey and Martha Chase, "Independent Functions of Protein and Nucleic Acid in Growth of Bacteriophage," *Journal of General Physiology* 36, no. 1 (1952): 39–56.

5. Erwin Chargaff, "Chemical Specificity of Nucleic Acids and Mechanism of Their Enzymatic Degradation," *Experientia* 6, no. 6 (1950): 201–209.

6. István Hargittai, *The DNA Doctor: Candid Conversations with James D. Watson* (Singapore: World Scientific, 2007), p. 60.

7. "Our Future Scientists (Panel Discussion)," *New York Academy of Sciences Magazine*, spring 2009, pp. 22–24; actual quote, p. 22.

8. Peter Medawar, "Lucky Jim," in *Pluto's Republic* (Oxford University Press, 1982), pp. 270–78; actual quote, p. 275.

9. Erwin Schrödinger, *What Is Life? The Physical Aspect of the Living Cell* (Cambridge: Cambridge University Press, 1944).

10. James D. Watson's letter of June 20, 2006, to the author.

11. Rita Levi-Montalcini, *In Praise of Imperfection: My Life and Work* (New York: Basic Books, 1988), p. 5.

12. James D. Watson, *The Double Helix: A Personal Account of the Discovery of the Structure of DNA* (New York: New American Library, 1968).

13. Two books have dealt with Rosalind Franklin's life and oeuvre: Anne Sayre, *Rosalind Franklin & DNA* (New York: Norton, 1975);

Brenda Maddox, *Rosalind Franklin: The Dark Lady of DNA* (London: HarperCollins, 2002).

14. See, e.g., Aaron Klug, "The Discovery of the Double Helix," *Journal of Molecular Biology* 335 (2004): 3–26.

15. See, e.g., the notes by Max Perutz and others in *Science*, June 27, 1969, 1537–38, following the publication of Watson's book *The Double Helix*.

16. James D. Watson, *Genes, Girls, and Gamow: After the Double Helix* (New York: Alfred A. Knopf, 2002), p. 10.

17. "If I have seen farther [than others] it is by standing upon the shoulders of giants." This well-known quotation is from Isaac Newton's letter to Robert Hook in 1675, but there is at least one forerunner version of it, by the first-century Roman poet Marcus Lucanus: "[P]igmies placed on the shoulders of giants see more than the giants themselves." See M. E. Lines, *On the Shoulders of Giants* (Bristol, UK, and Philadelphia: Institute of Physics, 1994), p. 1.

18. James D. Watson, "Growing Up in the Phage Group," in *Phage and the Origins of Molecular Biology*, expanded ed., ed. J. Cairns, G. S. Stent, and J. D. Watson (Cold Spring Harbor, NY: Cold Spring Harbor Laboratory Press, 1992), p. 242. See also Ernst Peter Fischer and Carol Lipson, *Thinking about Science: Max Delbrück and the Origins of Molecular Biology* (New York and London: W. W. Norton and Co., 1988), pp. 183–84.

19. Max Perutz, quoted in *Inspiring Science: Jim Watson and the Age of DNA*, ed. J. R. Inglis, J. Sambrook, and J. A. Witkowski (Cold Spring Harbor, NY: Cold Spring Harbor Laboratory Press, 2003), p. 71. This is in a brief section containing Perutz's farewell letter of January 14, 2002, to Watson, three weeks before Perutz's death.

20. James D. Watson, *Avoid Boring People (Lessons from a Life in Science)* (New York: Alfred A. Knopf, 2007), "Remembered Lessons," pp. 343–47.

21. James D. Watson, "Succeeding in Science: Some Rules of Thumb," *Science* 261, September 24, 1993, pp. 1812–13.

22. Charlotte Hunt-Grubbe, "The Elementary DNA of Dr. Watson," *Sunday Times*, October 14, 2007.

CHAPTER 2

1. István Hargittai, *Candid Science: Conversations with Famous Chemists*, ed. Magdolna Hargittai (London: Imperial College Press, 2000), "Gertrude B. Elion," pp. 54–71; actual quote, p. 71.

2. Arthur Kornberg, *The Golden Helix: Inside Biotech Ventures* (Sausalito, CA: University Science Books, 1995), p. 19.

3. Hargittai, *Candid Science*, p. 68.

4. Ibid., p. 69.

5. Burroughs Wellcome no longer figures among the names of pharmaceutical companies. It disappeared in the process of mergers and acquisitions. The original name continues its existence in the Burroughs Wellcome Fund, which is dedicated to supporting research and education in the biomedical sciences.

6. George H. Hitchings, speech at the Nobel Banquet, Stockholm, December 10, 1988.

7. Gösta Gahrton, presentation speech for the Nobel Prize in Physiology or Medicine, in *Nobel Lectures Physiology or Medicine 1981–1990*, ed. Jan Lindsten (Singapore: World Scientific, 1993), pp. 551–52; Joseph E. Murray, "The First Successful Organ Transplants in Man," in ibid., pp. 558–70.

CHAPTER 3

1. Paul C. Lauterbur, "All Science Is Interdisciplinary—from Magnetic Moments to Molecules to Men," Nobel lecture, in *Les Prix Nobel. The Nobel Prizes 2003*, ed. Tore Frängsmyr (Stockholm: Nobel Foundation, 2004), pp. 245–51.

2. István Hargittai and Magdolna Hargittai, *Candid Science VI: More Conversations with Famous Scientists* (London: Imperial College Press, 2006), "Peter Mansfield," pp. 216–37.

3. Ibid., p. 233.

4. Felix W. Wehrli, "The Origins and Future of Nuclear Magnetic

Resonance Imaging," *Physics Today* (June 1992): 34–42. See also "Letters" in the January 1993 issue of *Physics Today* in response to Wehrli's article by Jerome R. Singer (pp. 15 and 94); Herman Y. Carr (p. 94); Sidney Millman (pp. 94–95); Felix W. Wehrli (pp. 95–96).

5. Raymond Damadian, "Tumor Detection by Nuclear Magnetic Resonance," *Science* 171 (March 19, 1971): 1151–53.

6. Balazs Hargittai and István Hargittai, *Candid Science V: Conversations with Famous Scientists* (London: Imperial College Press, 2005), "Paul C. Lauterbur," pp. 454–79; actual quote, p. 456.

7. Ibid., p. 460.

8. P. C. Lauterbur, "Image Formation by Induced Local Interactions: Examples Employing Nuclear Magnetic Resonance," *Nature* 242 (March 16, 1973): 190–91.

9. Hargittai and Hargittai, *Candid Science V*, p. 474.

10. Peter Mansfield, "Snap-Shot MRI," Nobel lecture, in *Prix Nobel. The Nobel Prizes 2003*, ed. Tore Frängsmyr (Stockholm: Nobel Foundation, 2004), pp. 266–83.

11. Richard Ernst, "Foreword," in *Current Development in Solid State NMR Spectroscopy* (Vienna, Austria: Springer-Verlag, 2003).

12. Hargittai and Hargittai, *Candid Science VI*, pp. 233–35.

13. Ibid., p. 228.

14. Lauterbur, "Image Formation by Induced Local Interactions," p. 190.

CHAPTER 4

1. Linus Pauling, "Modern Structural Chemistry," in *Nobel Lectures Chemistry 1942–1962* (Singapore: World Scientific, 1999), pp. 429–37; actual quote, p. 437.

2. Linus Pauling, *The Nature of the Chemical Bond and the Structure of Molecules and Crystals: An Introduction to Modern Structural Chemistry*, 3rd ed. (Ithaca, NY: Cornell University Press, 1960; 1st ed., 1939; 2nd ed., 1940).

3. István Hargittai and Magdolna Hargittai, *Candid Science VI: More Conversations with Famous Scientists* (London: Imperial College Press, 2006), "Matthew Meselson," pp. 40–61; actual quote, pp. 58–59.

4. Their coworker Richard Marsh compared the two for me in a conversation in 1999 at Caltech.

5. W. Lawrence Bragg, John C. Kendrew, and Max F. Perutz, "Polypeptide Chain Configuration in Crystalline Proteins," *Proceedings of the Royal Society* 203A (1950): 321–57.

6. Linus Pauling and Robert B. Corey, "Two Hydrogen-Bonded Spiral Configurations of the Polypeptide Chain," *Journal of the American Chemical Society* 72 (1950): 5349; Linus Pauling, Robert B. Corey, and Herman R. Branson, "The Structure of Proteins: Two Hydrogen-Bonded Helical Configurations of the Polypeptide Chain," *Proceedings of the National Academy of Sciences of the U.S.A.* 37 (1951): 205–211.

7. Linus Pauling, "The Discovery of the Alpha Helix," *Chemical Intelligencer* 2, no. 1 (1996): 32–38. This was one of the last publications of Linus Pauling: Zelek S. Herman and Dorothy B. Munro, *The Publications of Professor Linus Pauling*, http://charon.girinst.org/~zeke.

8. Robert Olby, *The Path to the Double Helix: The Discovery of DNA* (New York: Dover Publications, 1974; 1994), p. 291.

9. Letter of February 18, 1948, from Linus Pauling to Robert Corey; Pauling Papers, individual correspondence, File C, 13.10, Oregon State University Library, quoted in Soraya de Chadarevian, *Designs for Life: Molecular Biology after World War II* (Cambridge: Cambridge University Press, 2002), p. 106.

10. Letter of February 25, 1948, from Robert Corey to Linus Pauling; ibid.

11. Dorothy C. Hodgkin and D. P. Riley, "Some Ancient History of Protein X-Ray Analysis," in *Structural Chemistry and Molecular Biology*, ed. Alexander Rich and N. Davidson (San Francisco and London: W. H. Freeman, 1968), pp. 15–28.

12. J. Desmond Bernal, "The Material Theory of Life," *Labour Monthly* (1968): 323–26.

13. István Hargittai, *Candid Science II: Conversations with Famous*

Biomedical Scientists, ed. Magdolna Hargittai (London: Imperial College Press, 2002), "Max F. Perutz," pp. 280–95; actual quote, p. 288.

14. Max Perutz, "Obituary: Linus Pauling," *Structural Biology* 1 (1994): 667–71.

15. Alan Mackay, "Two Times a Winner," *New Scientist*, November 25, 1995, pp. 52–53.

16. István Hargittai, *Candid Science: Conversations with Famous Chemists*, ed. Magdolna Hargittai (London: Imperial College Press, 2000), "The Great Soviet Resonance Controversy," pp. 8–13.

17. *Sostoyanie teorii khimicheskogo stroeniya v organicheskoi khimii* (The State of Affairs of the Theory of Chemical Structure in Organic Chemistry) (Moscow: Izdatel'stvo Akademii Nauk SSSR, 1952).

18. Hargittai, *Candid Science*, "Linus Pauling," pp. 2–7; actual quote, p. 5.

CHAPTER 5

1. Freeman J. Dyson, *The Sun, the Genome, and the Internet: Tools of Scientific Revolutions* (New York: Oxford University Press, 1999), p. 26.

2. István Hargittai, *Candid Science II: Conversations with Famous Biomedical Scientists*, ed. Magdolna Hargittai (London: Imperial College Press, 2002), "Frederick Sanger," pp. 72–83; actual quote, p. 74.

3. Ibid., p. 76.

4. Ibid.

5. Frederick Sanger, "Sequences, Sequences, and Sequences," *Annual Review of Biochemistry* 57 (1988): 1–28.

6. Ibid., p. 3.

7. The new tools were deemed so important that their discoverers were eventually awarded Nobel Prizes in Chemistry. Thus, Archer J. P. Martin and Richard L. M. Synge received one in 1952 for the so-called partition chromatography, and Stanford Moore and William H. Stein received another in 1972 for the so-called ion-exchange chromatography.

8. Arne Tiselius and Frederick Sanger, "Adsorption Analysis of Oxidized Insulin," *Nature* 160 (1947): 433–34.

9. Rodney R. Porter and Gerald M. Edelman shared the Nobel Prize in Physiology or Medicine in 1972 "for their discoveries concerning the chemical structure of antibodies." Porter was killed in an automobile accident in 1985.

10. Frederick Sanger, "The Chemistry of Insulin," in *Nobel Lectures: Chemistry 1942–1962* (Singapore: World Scientific, 1999), pp. 544–56; actual quote, p. 556.

11. Sanger, "Sequences, Sequences, and Sequences," pp. 11–14.

12. Hargittai, *Candid Science II*, p. 80.

13. Dyson, *The Sun, the Genome, and the Internet*, p. 30.

14. Catherine Brady, *Elizabeth Blackburn and the Story of Telomeres: Deciphering the Ends of DNA* (Cambridge, MA, and London: MIT Press, 2007), p. 34.

15. Sidney Altman, "MRC LMB—April 2003," talk to the MRC LMB meeting commemorating the fiftieth anniversary of the double-helix discovery, Cambridge, April 26, 2003. I am grateful to Sidney Altman for a copy of his talk.

16. Frederick Sanger's letter of January 26, 2002, to the author.

CHAPTER 6

1. Árpád Furka, "Combinatorial Chemistry," *Chemical Intelligencer* 5, no. 1 (1999), pp. 22–27; actual quote, p. 22.

2. István Hargittai, *Candid Science III: More Conversations with Famous Chemists*, ed. Magdolna Hargittai (London: Imperial College Press, 2003), "Bruce Merrifield," pp. 206–219.

3. R. Bruce Merrifield, *Journal of the American Chemical Society* 85 (1963): 2149–54.

4. Árpád Furka, "Combinatorial Chemistry: 20 Years On . . . ," *Drug Discovery Today* 7 (2002): 1–7, http://members.iif.hu/furka.arpad/.

5. Hargittai, *Candid Science III*, "Árpád Furka," pp. 220– 29.

6. Mamo Asgedom, "Peptidkeverékek szintézise és összetételük vizs-gálata" (Synthesis of Peptide Mixtures and the Investigation of Their Compositions) (Budapest: Eötvös Loránd University, Department of Organic Chemistry, 1987).

7. Á. Furka, F. Sebestyén, M. Asgedom, and G. Dibó, "Cornucopia of Peptides by Synthesis," *Abstracts of the 14th International Congress of Biochemistry* 5 (1988): 47; see also *Highlights of Modern Biochemistry, Proceedings of the 14th International Congress of Biochemistry*, vol. 5, ed. A. Kotyk (Utrecht, Netherlands: VSP Books, 1989), p. 47; Á. Furka, F. Sebestyén, M. Asgedom, and G. Dibó, "More Peptides by Less Labour," *Abstracts of the 10th Symposium of Medicinal Chemistry* (1988): 288.

8. Á. Furka, F. Sebestyén, M. Asgedom, and G. Dibó, "General Method for Rapid Synthesis of Multicomponent Peptide Mixture," *International Journal of Peptide and Protein Research* 37 (1991): 487–93.

9. Árpád Furka, *Combinatorial Chemistry: Principles and Techniques* (Budapest, 2007), http://members.iif.hu/furka.arpad/.

CHAPTER 7

1. Rosalyn Yalow, banquet speech at the Nobel Banquet, Stockholm, December 10, 1977.

2. Eugene Straus's excellent book on Yalow was published a few months later: *Rosalyn Yalow Nobel Laureate: Her Life and Work in Medicine. A Biographical Memoir* (New York and London: Plenum Trade, 1998).

3. István Hargittai, *Candid Science II: Conversations with Famous Biomedical Scientists*, ed. Magdolna Hargittai (London: Imperial College Press, 2002), "Rosalyn Yalow," pp. 518–23.

4. Rudolf Schoenheimer, *The Dynamic State of Body Constituents* (Cambridge, MA: Harvard University Press, 1942).

5. Magdolna Hargittai and István Hargittai, *Candid Science IV: Conversations with Famous Physicists* (London: Imperial College Press, 2004), "Maurice Goldhaber," pp. 214–31; actual quote, p. 221.

6. Hargittai and Hargittai, *Candid Science IV*, "Mildred Dresselhaus," pp. 546–69; actual quote, p. 549.

7. Ibid.

8. Ibid.

9. S. A. Berson, R. S. Yalow, A. Bauman, M. A. Rothschild, K. Newerly, "Insulin-I^{131} Metabolism in Human Subjects: Demonstration of Insulin Binding Globulin in the Circulation of Insulin-Treated Subjects," *Journal of Clinical Investigation* 35 (1956): 170–90.

10. Rosalyn S. Yalow, "Radioimmunoassay: A Probe for Fine Structure of Biologic Systems," in *Nobel Lectures: Physiology or Medicine 1971–1980* (Singapore: World Scientific, 1992), pp. 447–68; actual quote, p. 449.

11. Rosalyn S. Yalow, "Autobiography," in *Nobel Lectures: Physiology or Medicine 1971–1980* (Singapore: World Scientific, 1992), pp. 440–46; actual quote, p. 444.

12. Rolf Luft, "The Nobel Prize for Physiology or Medicine," ibid., pp. 357–59; actual quote, p. 358. Winston Churchill gave a speech to the House of Commons on August 20, 1940, which—in reference to the Royal Air Force—included the following phrase: "Never in the field of human conflict was so much owed by so many to so few." It is better known in a slightly abbreviated version: "Never was so much owed by so many to so few."

CHAPTER 8

1. Balazs Hargittai and István Hargittai, *Candid Science V: Conversations with Famous Scientists* (London: Imperial College Press, 2005), "Dan Shechtman," pp. 76–93; actual quote, p. 83.

2. Ibid., "Roger Penrose," pp. 36–55.

3. Martin Gardner, "Extraordinary Nonperiodic Tiling That Enriches the Theory of Tiles," *Scientific American* 236 (1977): 110–21.

4. Alan L. Mackay, "Crystallography and the Penrose Pattern," *Physica* 114A (1982): 609–613.

5. Hargittai and Hargittai, *Candid Science V*, "Dan Shechtman," pp. 76–93.

6. Alan L. Mackay, "A Dense Non-Crystallographic Packing of Equal Spheres," *Acta Crystallographica* 15 (1962): 916–18.

7. Private e-mail communication from Dan Shechtman, October 22, 2009.

8. D. Shechtman, I. Blech, D. Gratias, and J. W. Cahn, "Metallic Phase with Long Range Orientational Order and No Translational Symmetry," *Physical Review Letters* 53 (1984): 1951–53.

9. D. Levine and P. J. Steinhardt, "Quasicrystals: A New Class of Ordered Structures," *Physical Review Letters* 53 (1984): 2477–80.

10. István Hargittai and Magdolna Hargittai, *In Our Own Image: Personal Symmetry in Discovery* (New York: Kluwer Academic/Plenum Publishers, 2000), pp. 170–71.

11. From Max Planck, *A Scientific Autobiography* (1948), quoted in *Oxford Dictionary of Scientific Quotations*, ed. W. F. Bynum and Roy Porter (Oxford: Oxford University Press, 2005), p. 494.

12. See, e.g., Linus Pauling, "Interpretation of So-Called Icosahedral and Decagonal Quasicrystals of Alloys Showing Apparent Icosahedral Symmetry Elements as Twins of an 820-Atom Cubic Crystal," in *Symmetry 2: Unifying Human Understanding*, ed. István Hargittai (Oxford: Pergamon Press, 1989), pp. 337–39.

CHAPTER 9

1. Balazs Hargittai and István Hargittai, *Candid Science V: Conversations with Famous Scientists* (London: Imperial College Press, 2005), "Alan G. MacDiarmid," pp. 400–409; actual quote, p. 409.

2. Private e-mail communication of April 30, 2002, from Hideki Shirakawa to the author.

3. Takeo Ito, Hideki Shirakawa, and Sakuji Ikeda, "Simultaneous Polymerization and Formation of Polyacetylene Film on the Surface of Concentrated Soluble Ziegler-Type Catalyst Solution," *Journal of Polymer Science: Polymer Chemistry Edition* 12, no. 1 (1974): 11–20.

4. Hargittai and Hargittai, *Candid Science V*, "Alan J. Heeger," pp. 410–27; actual quote, p. 419.

5. Ibid., p. 405.

6. Hideki Shirakawa, "The Discovery of Polyacetylene Film: The Dawning of an Era of Conducting Polymers," in *Les Prix Nobel: The Nobel Prizes 2000* (Stockholm: Almquist & Wiksell International, 2001), pp. 217–26; actual quote, p. 225.

7. Hargittai and Hargittai, *Candid Science V*, pp. 415–16.

8. Shirakawa, "The Discovery of Polyacetylene Film," p. 219.

9. Ibid., pp. 225–26.

10. Hargittai and Hargittai, *Candid Science V*, p. 409.

CHAPTER 10

1. István Hargittai, *Candid Science: Conversations with Famous Chemists*, ed. Magdolna Hargittai (London: Imperial College Press, 2000), "F. Sherwood Rowland," pp. 448–65; actual quotes, pp. 464, 458.

2. G. E. Miller, P. M. Grant, R. Kishore, F. J. Steinkruger, F. S. Rowland, and V. P. Guinn, "Mercury Concentrations in Museum Specimens of Tuna and Swordfish," *Science* 175 (1972): 1121–22.

3. The authors also dealt with the mercury content in freshwater fish and found that freshwater fish accumulated mercury, too—and that might be a consequence of industrial pollutants.

4. James E. Lovelock, "Atmospheric Fluorine Compounds as Indicators of Air Movements," *Nature* 230 (1971): 379.

5. J. E. Lovelock, R. J. Maggs, and R. J. Wade, "Halogenated Hydrocarbons in and over the Atlantic," *Nature* 241 (1973): 194–96.

6. F. Sherwood Rowland, private communication, March 8, 2010. This is at slight variance from what both Rowland and Molina stated in their respective Nobel lectures in 1995. According to their lectures, Rowland presented a list of several projects, including the CFC problem, and Molina chose the CFC problem for his postdoctoral project.

7. John R. McNeill, *Something New Under the Sun: An Environ-*

mental History of the Twentieth-Century World (New York: W. W. Norton, 2001).

8. See, e.g., Michael D. Lemonick with Dick Thompson, "What Is Destroying the Ozone?" *Time* magazine, November 3, 1986, p. 80.

9. F. S. Rowland and Mario J. Molina, "Chlorofluoromethanes in the Environment," *Reviews of Geophysics and Space Physics* 13 (1975): 1–35.

10. W. Sullivan, "Tests Show Aerosol Gases May Pose Threat to Earth," *New York Times*, September 26, 1974.

11. Hargittai, *Candid Science*, p. 462.

12. Ibid.

13. Lydia Dotto and Harold Schiff, *The Ozone War* (Garden City, NY: Doubleday, 1978), p. 24.

14. See, for example, his latest book, James Lovelock, *The Vanishing Face of Gaia: A Final Warning* (New York: Basic Books, 2009).

15. Sharon Roan, *Ozone Crisis: The 15-Year Evolution of a Sudden Global Emergency* (New York: Wiley & Sons, 1989), p. 43.

CHAPTER 11

1. Kary B. Mullis, *Dancing Naked in the Mind Field* (New York: Pantheon Books, 1998), p. 4.

2. Mullis, *Dancing Naked*, front matter.

3. Alfred B. Nobel's Will in Paris, November 27, 1895; see, e.g., István Hargittai, *The Road to Stockholm: Nobel Prizes, Science, and Sciences* (Oxford: Oxford University Press, 2002), p. 3.

4. István Hargittai, *Candid Science II: Conversations with Famous Biomedical Scientists*, ed. Magdolna Hargittai (London: Imperial College Press, 2002), "Kary B. Mullis," pp. 182–95.

5. Ibid., p. 185.

6. F. R. Stannard, "Symmetry of the Time Axis," *Nature* 211 (1966): 693–95.

7. Kary B. Mullis, "The Cosmological Significance of Time Reversal," *Nature* 218 (1968): 663–64.

8. István Hargittai, "*Nature*'s Maddox," *Chemical Intelligencer* 5, no. 2 (1999): 53–55; actual quote, p. 54.

9. Kary B. Mullis, "The Polymerase Chain Reaction," in *Nobel Lectures Chemistry 1991–1995*, ed. Bo G. Malström (Singapore: World Scientific, 1997), pp. 103–113.

10. Arthur Kornberg, *The Golden Helix: Inside Biotech Venture* (Sausalito, CA: University Science Books, 1995), p. 237.

11. R. K. Saiki, S. Scharf, F. Faloona, K. M. Mullis, G. T. Horn, A. A. Erlich, and N. Arnheim, "Enzymatic Amplification of Beta-Globin Genomic Sequences and Restriction Site Analysis for Diagnosis of Sickle-Cell Anemia," *Science* 230 (1985): 1350–54.

12. Kary B. Mullis and Fred A. Faloona, "Specific Synthesis of DNA in Vitro via a Polymerase-Catalyzed Chain Reaction," *Methods in Enzymology* 155 (1987): 355–50.

13. K. Mullis, F. Faloona, S. Scharf, R. Saiki, G. Horn, and H. Erlich, "Specific Enzymatic Amplification of DNA in Vitro: The Polymerase Chain Reaction," *Cold Spring Harbor Laboratory Symposium on Quantitative Biology* 51 (1986): 263–73 .

14. Hargittai, *Candid Science II*, p. 191.

15. Ibid., p. 184.

16. Ibid., p. 192.

17. Ibid.

CHAPTER 12

1. István Hargittai, *Candid Science III: More Conversations with Famous Chemists*, ed. Magdolna Hargittai (London: Imperial College Press, 2003), "Neil Bartlett," pp. 28–47; actual quote, p. 38.

2. Neil Bartlett, "Xenon Hexafluoroplatinate(V) $Xe^+[PtF6]^-$," *Proceedings of the Chemical Society* (June 1962): 218.

3. Ibid.

4. Hargittai, *Candid Science III*, pp. 28–47, and e-mail exchanges with Neil Bartlett in 2000 and 2001.

5. Howard H. Claassen, Henry Selig, and John G. Malm, "Xenon Tetrafluoride," *Journal of the American Chemical Society* 84 (1962): 3593.

6. R. Hoppe, W. Dähne, H. Mattauch, and K. M. Rödder, "Fluorierung von Xenon," *Angewandte Chemie* 74 (1962): 903.

7. J. Slivnik, B. Brcic, B. Volavsek, J. Marsel, V. Vrscaj, A. Smalc, B. Frlec, and Z. Zemljic, "Über die Synthese von XeF$_6$," *Croatica Chemica Acta* 34 (1962): 253.

8. P. R. Fields, L. Stein, and M. H. Zirin, "Radon Fluoride," *Journal of the American Chemical Society* 84 (1962): 4164–65.

9. Private e-mail communication from Neil Bartlett, June 13, 2000.

10. Don M. Yost and Albert L. Kaye, "An Attempt to Prepare a Chloride or Fluoride of Xenon," *Journal of the American Chemical Society* 55 (1933): 3890–92.

11. Ibid.

12. Don M. Yost, "A New Epoch in Chemistry," in *Noble-Gas Compounds*, ed. Herbert H. Hyman (Chicago and London: University of Chicago Press, 1963), pp. 21–22; actual quote, p. 21.

13. Ibid.; actual quote, p. 22.

14. Ibid.

15. Ibid.

16. Hargittai, *Candid Science III*, p. 47.

17. Leonid Khriachtchev, Mika Pettersson, Nino Runeberg, Jan Lundell, and Markku Räsänen, "A Stable Argon Compound," *Nature* 406 (August 24, 2000): 874–76.

18. Ronald J. Gillespie and István Hargittai, *The VSEPR Model of Molecular Geometry* (Boston: Allyn & Bacon, 1991).

19. L. Graham, O. Graudejus, N. K. Jha, and N. Bartlett, "Concerning the Nature of XePtF$_6$," *Coordination Chemistry Review* 197 (2000): 321–34.

20. N. Bartlett, "Forty Years of Fluorine Chemistry," in *Fluorine Chemistry at the Millennium*, ed. R. E. Banks (Amsterdam: Elsevier, 2000), pp. 29–55.

21. Michael Freemantle, "Chemistry at Its Most Beautiful," *Chemical and Engineering News* (August 25, 2003): 27–30; actual quote, p. 27.

22. Primo Levi, *The Periodic Table*, transl. from the Italian by Raymond Rosenthal (New York: Schocken Books, 1984), p. 4.

CHAPTER 13

1. William Lanouette with Bela Silard, *Genius in the Shadows: A Biography of Leo Szilard, The Man Behind the Bomb* (Chicago: University of Chicago Press, 1994), p. 441.

2. H. H. Goldstine, *The Computer from Pascal to von Neumann* (Princeton, NJ: Princeton University Press, 1972), pp. 279–80.

3. C. P. Snow, *Variety of Men* (New York: Charles Scribner's Sons, 1967), "Rutherford," pp. 3–20; actual quote, p. 16.

4. S. R. Weart and Gertrud Weiss Szilard, eds., *Leo Szilard: His Version of the Facts. Selected Recollections and Correspondence* (Cambridge, MA: MIT Press, 1978), p. 17.

5. Ernest Rutherford was quoted in a report "Atomic Transmutation," in *Nature* 132 (1933): 432–33.

6. Snow, *Variety of Men*, p. 13.

7. István Hargittai, *The Martians of Science: Five Physicists Who Changed the Twentieth Century* (New York: Oxford University Press, 2006), pp. 188–95.

8. Weart and Weiss Szilard, *Leo Szilard*, p. 54.

9. Ibid., pp. 189–92.

10. Ibid., p. 163.

11. István Hargittai, *Candid Science II: Conversations with Famous Biomedical Scientists*, ed. Magdolna Hargittai (London: Imperial College Press, 2002), "François Jacob," pp. 84–97; actual quote, p. 91.

12. Jacques Monod, "From Enzymatic Adaptation to Allosteric Transitions," in *Nobel Lectures Physiology or Medicine 1963–1970* (Singapore: World Scientific, 1999), p. 199. François Jacob, André Lwoff, and Jacques Monod were awarded the Nobel Prize in Physiology or Medicine in 1965 "for their discoveries concerning genetic control of enzyme and virus synthesis."

13. Istvan Hargittai, *Judging Edward Teller: A Closer Look at One of the Most Influential Scientists of the Twentieth Century* (Amherst, NY: Prometheus Books, 2010), pp. 213–14.

14. Ibid., p. 337.

15. Alice Calaprice, ed. and collector, *The Expanded Quotable Einstein* (Princeton, NJ: Princeton University Press, 2000), p. 175; from "Atomic War or Peace," *Atlantic Monthly*, November 1945.

16. Letter of May 27, 1960, from John F. Kennedy to Leo Szilard, addressed to the Memorial Hospital at 68th Street and York Avenue. I thank the late George Marx, Budapest, for a copy of this letter.

CHAPTER 14

1. Magdolna Hargittai and István Hargittai, *Candid Science IV: Conversations with Famous Physicists* (London: Imperial College Press, 2004), "Freeman J. Dyson," pp. 440–77; actual quote, p. 454.

2. Interview with Edward Teller by an unidentified interviewer for the Academy of Achievement, Palo Alto, California, September 30, 1990.

3. Luis W. Alvarez, *Alvarez: Adventures of a Physicist* (New York: Basic Books, 1987), p. 166.

4. John A. Wheeler, *Geons, Black Holes & Quantum Foam: A Life in Physics*, with Kenneth Ford (New York: W. W. Norton, 2000), p. 207.

5. See, e.g., Wheeler, *Geons*, p. 206, n. 5.

6. Edward Teller's memorandum in George A. (Jay) Keyworth's office on September 20, 1979.

7. S. M. Ulam, *Adventures of a Mathematician* (Berkeley, CA: University of California Press, 1991), p. 220.

8. Private e-mail communication from Richard Garwin, March 25, 2009.

9. Edward Teller, "A New Thermonuclear Device," LA-1230, Los Alamos Scientific Laboratory of the University of California, April 4, 1951.

10. Hans A. Bethe, "Comments on the History of the H-bomb," *Los Alamos Science*, Fall 1982, pp. 43–53; actual quote, p. 49.

11. Letter in mid-winter 1949 from Edward Teller to Maria Goeppert Mayer; UCSD Library/Hoover Archives (No. 3 in 1949).

12. Istvan Hargittai, *Judging Edward Teller: A Closer Look at One of the Most Influential Scientists of the Twentieth Century* (Amherst, NY: Prometheus Books, 2010), pp. 239–40.

13. Transcripts of the group interview with Teller at Los Alamos National Laboratory, June 1993, by Charles R. (Chuck) Hansen. I am grateful to Richard Garwin for the transcripts of the interview.

14. In referring to Szilard and his Hungarian colleagues asking Albert Einstein to warn President Roosevelt in 1939 about a possible German atomic bomb, Isidor Rabi stated, "The Germans owed a lot to Szilard." According to Rabi, other channels—outside the government—would have been more efficient in starting the American atomic bomb project.

CHAPTER 15

1. Balazs Hargittai and István Hargittai, *Candid Science V: Conversations with Famous Scientists* (London: Imperial College Press, 2005), "Vera C. Rubin," pp. 246–65; actual quote, p. 251.

2. Quoted in Eamon Harper, "In Appreciation George Gamow: Scientific Amateur and Polymath," *Physics in Perspective* 3 (2001): 335–72; actual quote, p. 369.

3. George Gamow, *The Creation of the Universe* (New York: Viking Press, 1952), p. vii.

4. Ibid.

5. Ibid.

6. See, e.g., Yakov B. Zeldovich, *My Universe: Selected Reviews* (Routledge, 1992), p. 3.

7. Gennadi A. Sardanashvily (Г. А. Сарданашвили), *Дмитрий Иваненко—суперзвезда советской физики. Ненаписанные мемуары* (Dmitri Ivanenko—Superstar of Soviet Physics. Unwritten Memoirs) (Moscow: URSS Publisher, 2010), p. 209.

8. The original Russian was translated into English and published in

2002: G. Gamow, D. Ivanenko, and L. Landau, "World Constants and Limiting Transition," *Physics of Atomic Nuclei* 65 (2002): 1373–75.

9. Sardanashvily, *Dmitri Ivanenko*, p. 215.

10. Eddington was so sure of the fundamental correctness of his suggestion that he was willing to ignore the fact that the temperature might be insufficiently high in the interior of stars. In time, it was understood that (just like Gamow's discovery of the possibility of alpha decay) there are tunneling effects that make this reaction feasible. Eddington's charging ahead with his theory even though it had not yet fit together with observations reminds us of Pauling's optimism about the correctness of his alpha-helix model of proteins at the time when some experimental facts showed a discrepancy with the model (see chapter 4).

11. George Gamow, "Nuclear Transformations and the Origin of the Chemical Elements," *Ohio Journal of Science*, no. 5 (September 1935): 406–413.

12. See, e.g., David N. Schramm, "Cosmological Implications of Light Element Abundances: Theory," *PNAS USA* 90 (1993): 4782–88.

13. Peter Garrity, *The Galloping Gamows* (private ed., 2007), p. 63. Referring to Robert Herman, another member of the group investigating the origin of the universe, Gamow jokingly lamented that he "stubbornly refuses to change his name to Delter." (Gamow, *The Creation of the Universe*, p. 65.)

14. Ralph A. Alpher, Hans A. Bethe, and George Gamow, "The Origin of Chemical Elements," *Physical Review* 73 (1948): 803–804.

15. Hargittai and Hargittai, *Candid Science V*, p. 250.

16. George Gamow, "Expanding Universe and the Origin of Galaxies," *Det Kongelige Danske Videnskabernes Selskab, Matematisk–fysiske Meddelelser* 27, no. 10 (1953): 1–15.

17. Magdolna Hargittai and István Hargittai, *Candid Science IV: Conversations with Famous Physicists* (London: Imperial College Press, 2004), "Arno Penzias," pp. 272–85; actual quote, pp. 274–75.

18. Arno A. Penzias, "The Origin of Elements," in *Nobel Lectures in Physics 1971–1980* (Singapore: World Scientific, 1992), pp. 444–62.

19. Hargittai and Hargittai, *Candid Science IV*, p. 277; see also "Robert W. Wilson" in the same volume, pp. 288–89.

20. Ralph A. Alpher and Robert Herman, *Genesis of the Big Bang* (New York: Oxford University Press, 2001), p. 121.

21. George Gamow, *My World Line: An Informal Autobiography* (New York: Viking Press, 1970).

INDEX

Charterhouse Library

67893